TECHNOLOGIES FOR CMOM ACTIVITIES IN WASTEWATER COLLECTION SYSTEMS

WEF Special Publication

2022

Water Environment Federation
601 Wythe Street
Alexandria, VA 22314-1994 USA
www.wef.org

ISBN: 978-1-57278-424-6

About WEF

The Water Environment Federation (WEF) is a not-for-profit technical and educational organization of 35,000 individual members and 75 affiliated Member Associations representing water quality professionals around the world. Since 1928, WEF and its members have protected public health and the environment. As a global water sector leader, our mission is to connect water professionals; enrich the expertise of water professionals; increase the awareness of the impact and value of water; and provide a platform for water sector innovation. To learn more, visit www.wef.org.

Prepared by the Technologies for CMOM Activities in Wastewater Collection Systems Task Force of the Water Environment Federation

Jane McLamarrah, PE, PhD, ENV SP, *Chair*

Ray Bahr, PE
Chein-Chi Chang, PhD, PE
Anjulie Cheema
Alex Churchill
Steve Donovan
Kevin Draganchuk, PE, BCEE
Mark Draper, PE
Rudy Fernandez, PE
Mayra Giraldo, EIT
David Hamberlin
Stacy N. Hayden, PE
Phil Hubbard, PE
Reese Johnson, PE, PMP
Brock A. Jones, PE
Christene Mitchell, PE
Susan Moisio, PE
Daniel J. Murray Jr., PE, BCEE
Dan O'Leary
Raghavender Nednur, PE
Christopher Ranck, PE, BCEE, D. WRE, ENV SP
Diane Ryu
Derek Schwanke, PE
Charles H. Starling III, PE, BCEE
Jimmy Stewart, Collection Systems Committee Chair
Dianne Sumego, PE
Tim Sumner, PE
Matthew Van Doren, PE
Christy Willson, PE
Patrick L. Woodall

Under the Direction of the **Collection Systems Subcommittee** of the **Technical Practice Committee**

Special Publications of the Water Environment Federation

The WEF Technical Practice Committee (formerly the Committee on Sewage and Industrial Wastes Practice of the Federation of Sewage and Industrial Wastes Associations) was created by the Federation Board of Control on October 11, 1941. The primary function of the Committee is to originate and produce, through appropriate subcommittees, special publications dealing with technical aspects of the broad interests of the Federation. These publications are intended to provide background information through a review of technical practices and detailed procedures that research and experience have shown to be functional and practical.

Contents

List of Figures

List of Tables

Preface

Many wastewater collection system operators have successfully adopted capacity, management, operations, and maintenance (CMOM) activities to move from reactive operation and maintenance (O&M) to more proactive O&M activities. This publication identifies technologies that can be applied to further this movement into more preventive and even predictive O&M activities and continue to reduce reactive O&M. The identified technologies include equipment or systems that:

- Determine if wastewater collection system assets are functioning properly;
- Assess the physical condition of various types of such assets;
- Repair or rehabilitate the condition of such assets; and
- Improve the operation and maintenance of such assets.

Several of the identified technologies are becoming more cost-effective to implement by both small utilities with limited budgets and by large utilities that may need significant quantities of the equipment. Other technology-focused processes are becoming easier to implement with existing operational skill levels or now have more availability of personnel with the required skills.

The publication is targeted to supervisors, operators, and maintenance personnel. The increased use of innovative technology, smart data, system optimization, and real-time controls will enhance optimization of O&M budgets and contribute to an overall reduction in life-cycle costs.

This first edition of this publication was produced under the direction of Jane McLamarrah.

Authors' and reviewers' efforts were supported by the following organizations:

AECOM
Black & Veatch
CEA Engineers, P.C.
City of Atlanta - Department of Watershed Management
Crawford, Murphy & Tilly
dms water solutions co., llc
Emerald Coast Utilities Authority
Garver LLC

Green Mountain Pipeline Services
Hampton Roads Sanitation District
Horner & Shifrin, Inc.
Hydromax USA
Infosense
Jacobs
Kimley-Horn
Metropolitan Sewer District of Greater Cincinnati
Stantec
U.S. Environmental Protection Agency, Office of Research and Development
Water Environment Association of Texas (WEAT) Collection Systems Committee
Zhejiang University of Water Resources and Electric Power

1

Introduction

Jane McLamarrah, PE, PhD, ENV SP & Mayra Giraldo, EIT

1.0 PURPOSE

The purpose of this Special Publication is to provide guidance and direction to those individuals responsible for the operation and maintenance (O&M) of wastewater collection systems for the increasing use of new or innovative technology, smart data, system optimization, and real-time controls. The greater adoption of asset management and capacity, management, operations, and maintenance (CMOM) principles for wastewater collection systems since the late 1990s and early 2000s has increased the emphasis on providing a high level of service to customers, has reduced regulatory noncompliance events, and has lessened the risks associated with emergency asset failures. These desirable utility benefits were achieved largely through a shift from "reactive" to "proactive" O&M activities that often led to savings in avoided costs of overtime; emergency repairs and construction; remediation efforts; lawsuits; and insurance premiums. This Special Publication is intended to enhance such utility benefits by identifying and describing existing, innovative, or new technologies that can be implemented to shift even further from proactive to "preventive" and "predictive" O&M activities.

Although asset management and CMOM principles are important considerations in themselves, the focus of this Special Publication will be on the operational aspects of implementing technologies to improve the effectiveness and efficiency of asset performance and asset repair, rehabilitation, or replacement. Only brief introductory discussion will be included for the "why" and "how" these technologies are, or should be, implemented. Readers requiring a more detailed discussion of management, policy, and regulatory issues are directed to the companion publication *Wastewater Collection Systems Management: Water Environment Federation (WEF) Manual of Practice* No. 7 (WEF, 2021).

2.0 TARGET AUDIENCE

This Special Publication is directed primarily at wastewater collection system operators and maintenance personnel although many others may also benefit. Upper and senior managers as well as scientific and research staff will benefit from the identification of potential technologies and the protocols and processes for managing data and are further directed to the *WEF Manual of Practice No. 7* companion publication. Engineering and design staff will identify potential innovative technological solutions to manage asset life cycles and costs, and are further directed to the companion publication *Existing Sewer Evaluation and Rehabilitation: WEF Manual of Practice No. FD-6, American Society of Civil Engineers (ASCE) Manuals and Reports on Engineering Practice No.* 62 (WEF et al., 2009).

3.0 IMPLEMENTING CONTINUOUS IMPROVEMENT

The importance of applying proper asset management and CMOM principles to wastewater collection system planning, design, installation, and management has been documented in many publications by both regulatory agencies and professional organizations. The reader is directed to publications such as *Effective Utility Management: A Primer for Water and Wastewater Utilities* (U.S. EPA et al., 2017) and *Guide for Evaluating Capacity, Management, Operation, and Maintenance (CMOM) Programs at Sanitary Sewer Collection Systems* (U.S. EPA, 2005) as well as to our companion *WEF Manual of Practice No.* 7 (WEF, 2021).

Implementation of a continuous improvement program is a key element of many of the recommended asset management and CMOM principles. Figure 1.1 illustrates the iterative step process of "plan-do-check-act" or "plan-do-check-adjust" used to implement continuous improvement.

3.1 Reactive to Proactive

For many years, wastewater collection systems were operated with a reactive maintenance strategy where assets were "run-to-failure" with little maintenance until problems arose. This approach typically led to situations where sanitary sewer overflows (SSOs) and dry weather combined sewer overflows (CSOs) were a routine occurrence that were only reported to regulatory agencies when large volumes were discharged or a "fish kill" event occurred. As asset management and CMOM principles began to be

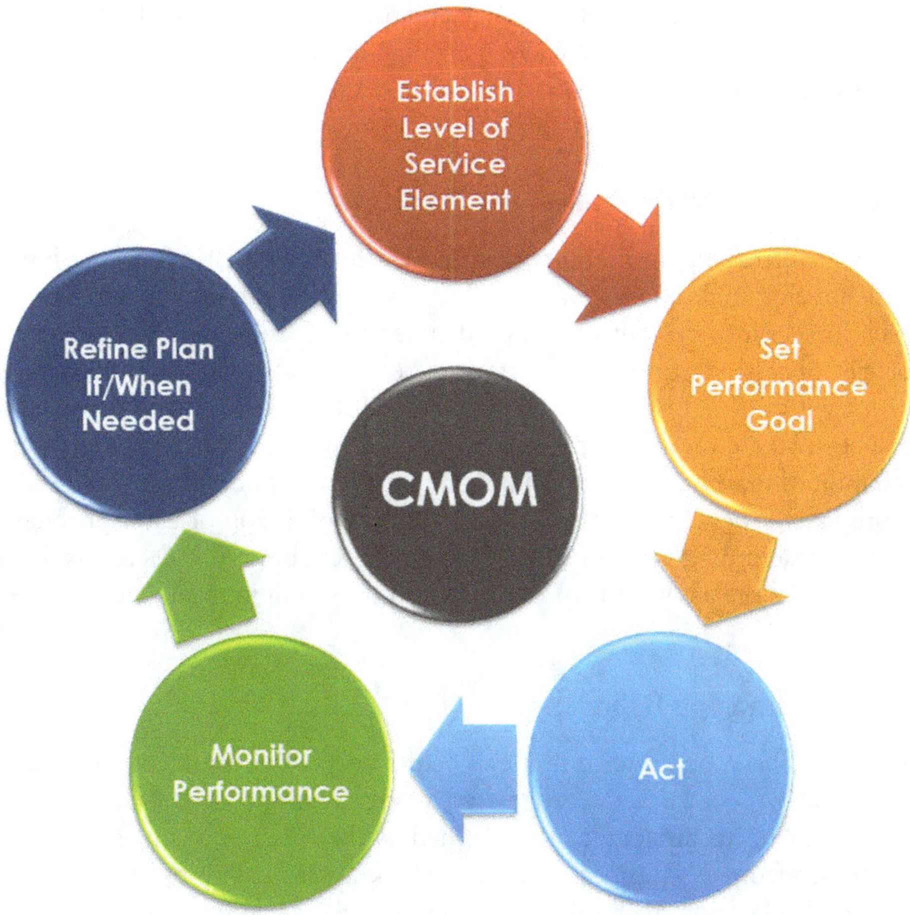

FIGURE 1.1 Continuous Improvement Schematic

applied in the late 1990s and early 2000s, utilities increasingly adopted a more continuous improvement approach either voluntarily or dictated by regulatory consent decrees and National Pollutant Discharge Elimination System (NPDES) permits.

Under the continuous improvement approach, level of service was established, associated performance goals were defined, and programs were designed to meet the goals. Information was collected and tracked on how the programs were meeting performance goals, especially SSO and CSO reduction goals, and whether overall system efficiency was improving. Periodically, wastewater collection system protocols, practices, procedures, activities, and programs were reviewed and adjusted to better meet level of service defined performance goals. The emphasis on meeting performance goals, and continually raising the performance bar of those goals, allowed operators to spend more time, resources, and budgets on addressing deteriorating assets rather on already failed assets and thus reduce, and even prevent, overflows. See Section 6.0, Suggested Readings, for additional information on establishing level of service and performance goals. The American Water Works Association (AWWA), through its annual AWWA Benchmarking Survey (AWWA, 2018), assists utilities with monitoring performance goals for commonly used goals.

3.2 Moving to Preventive and Predictive

With the increasing availability, and more affordable cost, of new or innovative technologies, wastewater collection systems are now able to move even further into preventive and predictive O&M. Technologies can ensure proper operation of assets and transmit real-time alarms when the asset is operating outside of specified parameters. More specific and targeted data can be acquired on the physical condition of assets to be able to base repair, rehabilitation, or renewal decisions on actual condition data for more and more asset types. These improvements allow prediction of asset problems and corresponding "just in time" repair and rehabilitation decisions, thus maximizing the asset's return on investment and minimizing asset replacement costs.

3.3 Technology Definition

Technology is a broad term that can be defined as the application of scientific knowledge for practical purposes, especially in industry. When applied to equipment, technology can be defined as the machinery and equipment developed from the application of scientific knowledge. For purposes of this Special Publication, technologies are considered as equipment or systems (but not methods like open cut replacement) that:

- Determine if wastewater collection system assets are functioning properly,
- Assess the physical condition of various types of such assets,
- Repair or rehabilitate the condition of such assets, and
- Improve the operation or maintenance of such assets.

This Special Publication concentrates on the new or innovative technologies that may not yet be familiar to O&M personnel. Though this Special Publication is not a complete body of knowledge on any of the topics, additional references and lists of further reading materials are provided at the end of each chapter.

3.4 Intelligent Systems

The greater use of technologies requires corresponding increases in associated data management practices and protocols necessary to acquire valid and accurate data required to use technologies, and the information acquired from technologies, effectively. Operation and maintenance personnel must understand what can be accomplished with available technologies and must be properly trained in the retrieval and analysis of performance and condition data from these technologies. Further, utilities and all personnel levels must expend significant investment and personnel training in cyber-security systems to protect their data and all remote-control systems. Protocols and practices must be implemented to facilitate proper data management, including data storage, security, and retrieval; data availability and sharing; visualization and ease of use of data; regulatory notifications if required; and public or customer transparency considerations. As utilities adopt these systems, the resulting "intelligent wastewater collection systems" will reduce overflows, operate more effectively, and use customer revenues more cost-effectively.

4.0 SPECIAL PUBLICATION ORGANIZATION

This Special Publication is organized in the following chapters:

- Chapter 2, "Wastewater Collection System Asset Inventories and Screening Inspections," will focus on wastewater collection system asset location and inventory; asset monitoring; and preliminary visual or screening inspection technologies and practices to determine when further condition assessments are required.
- Chapter 3, "Collection System Operation and Maintenance-Related Technologies," will focus on technologies to better track, and thus

prevent, sewer system stoppages and overflows with a focus on advance warning and predictive flow analyses, along with emergency response and corrective actions to improve response time and effective overflow control.

- Chapter 4, "Asset Condition Assessment," will detail the types of asset condition assessment technologies for gravity sewers and pressurized pipes.
- Chapter 5, "Collection System Rehabilitation and Replacement Technologies," will describe collection system rehabilitation and replacement technologies.
- Chapter 6, "Intelligent Wastewater Collection Systems," will identify trends in applying intelligent data and equipment through the implementation of more sophisticated controls and technologies, including a discussion of staffing skills required to install, operate, and maintain these increasingly automated, sensitive, or expensive equipment installations.

5.0 REFERENCES

American Water Works Association. (2018). *2018 AWWA utility benchmarking: Performance management for water and wastewater.* https://www.awwa.org/Portals/0/files/publications/documents/BenchmarkingLookInside.pdf

U.S. Environmental Protection Agency. (2005). *Guide for evaluating capacity, management, operation, and maintenance (CMOM) programs at sanitary sewer collection systems* (EPA 305-B-05-002). https://www3.epa.gov/npdes/pubs/cmom_guide_for_collection_systems.pdf

U.S. Environmental Protection Agency, Association of Metropolitan Water Agencies, Association of Clean Water Administrators, Association of State Drinking Water Administrators, American Public Works Association, American Water Works Association, National Association of Clean Water Agencies, National Association of Water Companies, Water Environment Federation, Water Environment & Reuse Foundation, & Water Research Foundation. (2017). *Effective utility management: A primer for water and wastewater utilities.* U.S. Environmental Protection Agency, Office of Wastewater Management. https://www.epa.gov/sites/production/files/2017-01/documents/eum_primer_final_508-january2017.pdf

Water Environment Federation. (2021). *Wastewater collection systems management* (6th ed., WEF Manual of Practice No. 7).

Water Environment Federation, American Society of Civil Engineers, & Environmental & Water Resources Institute. (2009). *Existing sewer evaluation and rehabilitation* (3rd ed., WEF Manual of Practice No. FD-6, ASCE Manuals and Reports on Engineering Practice No. 62). Water Environment Federation.

6.0 SUGGESTED READINGS

American Water Works Association. (2019). *Key data to inform government asset management policies.* https://www.awwa.org/Portals/0/AWWA/ETS/Research/KeyDataReport2019.pdf?ver=2019-09-13-133310-377

Envirosight LLC. (2018). Understanding CMOM. *Municipal Sewer & Water, 6*(4). https://www.mswmag.com/online_exclusives/2018/06/understanding-cmom_sc_00125

Georgia Association of Water Professionals. (2016). *Wastewater collection systems best management practices.* Georgia Association of Water Professionals Collection Systems Committee. https://cdn.ymaws.com/www.gawp.org/resource/resmgr/Collections_BMP_Manual/WW_Collection_Systems_BMP_Ma.pdf

Tscheikner-Gratl, F., Caradot, N., Cherqui, F., Leitão, J. P., Ahmadi, M., Langeveld, J. G., Le Gat, Y., Scholten, L., Roghani, B., Rodríguez, J. P., Lepot, M., Stegeman, B., Heinrichsen, A., Kropp, I., Kerres, K., Céu Almeida, M., Bach, P. M., Moy de Vitry, M., Marques, A. S., . . . Clemens, F. (2020). Sewer asset management: State of the art and research needs. *Urban Water Journal, 16*(9), 662–675. https://www.tandfonline.com/doi/pdf/10.1080/1573062X.2020.1713382?needAccess=true

U.S. Environmental Protection Agency. (2007). *Computer tools for sanitary sewer system capacity analysis and planning* (EPA 600/R-07/111). Office of Research and Development. https://nepis.epa.gov/Adobe/PDF/P1008BBP.pdf

U.S. Environmental Protection Agency. (2008). *Review of sewer design criteria and RDII predictions* (EPA 600/R-08/010). Office of Research and Development. https://nepis.epa.gov/Adobe/PDF/P1008BP3.pdf

U.S. Environmental Protection Agency. (2012). *SSOAP toolbox enhancements and case study* (EPA 600/R-12/690). Office of Research and Development. https://nepis.epa.gov/Adobe/PDF/P100G00U.pdf

U.S. Environmental Protection Agency. (2017). *Asset management programs for stormwater and wastewater systems: Overcoming barriers to development and implementation.* Office of Science and Technology, and U.S.

Environmental Protection Agency Region 9. https://www.epa.gov/sites/default/files/2018-01/documents/overcoming-barriers-to-development-and-implementation-of-asset-management-plans.pdf

U.S. Environmental Protection Agency Region 1. (2009). *Template for developing sewer collection system preventive maintenance and sewer overflow response plans.* https://www3.epa.gov/region1/sso/other/PreventiveMaintenanceSewerOverflowResponsePlanTemplateNovember2009.doc

Water Effective Utility Management. (n.d.). *Ten attributes of effectively managed water sector utilities.* https://www.watereum.org/resources/interactive-primer/ten-attributes/

Water Environment Federation Disinfection and Public Health Committee Waterborne Infectious Disease Outbreak Control Working Group. (2021). Translating wastewater surveillance data. *Water Environment and Technology, 33*(4).

Water Research Foundation. (n.d.). *Sustainable infrastructure management program learning environment.* https://simple.waterrf.org/

2

Wastewater Collection System Asset Inventories and Screening Inspections

Patrick L. Woodall, Alex Churchill, & Diane Ryu

1.0 ASSET INVENTORY

Maintaining an accurate collection system asset inventory is foundational to developing a well-designed continuous improvement program. Understanding the different components and materials that make up such a complex underground system critically affects the ability to develop effective maintenance protocols, make long-term asset rehabilitation decisions, and select appropriate technologies for predictive operations and maintenance (O&M).

The composition of assets within a specific collection system depends on variables such as the local geology, topography, age of the system, and size of population served. A comprehensive asset inventory incorporates data regarding each significant asset's location, function, and key attributes. Although asset data can be maintained on paper or in a spreadsheet or similar electronic format, many collection systems today have selected a geographic information system (GIS) to store and map asset information. Using a GIS system is considered the best practice, as they can greatly improve operational efficiency, information accessibility, analysis, reporting, and record keeping through visual representation of assets. The following sections of this chapter describe the primary assets found in collection systems and preliminary screening tools that can assist in guiding and prioritizing their maintenance.

1.1 Manholes

Manholes, also referred to as maintenance holes, allow operators and equipment to access sewer lines. Manholes are located when there is a change in grade, alignment, or size of sewer and at appropriate intervals to allow access for sewer line maintenance and inspection activities. Properly conducted manhole inspections help to ascertain the materials used in construction, identify potential sources of inflow and infiltration, identify structural defects, and provide information to identify past or present surcharge events. Examples of typical components for manholes are illustrated in Figure 2.1. See *Manhole Inspection and Rehabilitation* (Hughes, 2009) and *Developing a Manhole or Catch Basin Numbering System* (U.S. Environmental Protection Agency [U.S. EPA] Region 1, 2017) for manhole inspection, rehabilitation, and numbering systems.

1.2 Sewer Mains

Collection system sewer mains convey wastewater to a water resource recovery facility (WRRF), also known as water resource recovery centers or wastewater treatment plants. The two primary types of sewer systems in North America are separate sewer systems and combined sewer systems.

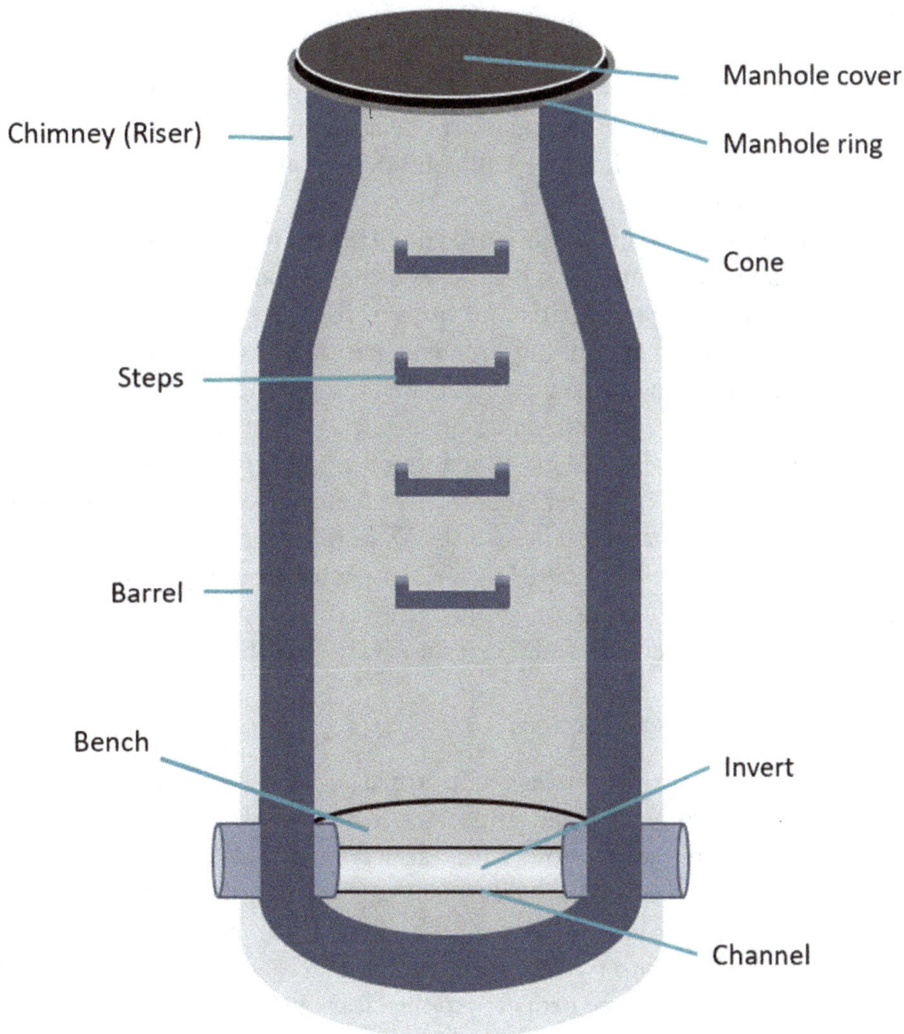

FIGURE 2.1 Components of a Manhole

Both systems convey wastewater by either gravity or pressure. Separate sewer systems collect stormwater and sanitary sewage in separate pipe networks. Water resulting from precipitation is discharged by the stormwater system to local streams or bodies of water. Sanitary sewage is conveyed to a WRRF for treatment.

Combined sewer systems were historically designed to collect stormwater and sanitary sewage in the same pipe and are found primarily in the downtown areas of older cities. In most cases, water in a combined sewer system is carried to a WRRF for treatment. However, in the case of excessive precipitation, the system capacity may be exceeded, resulting in discharge of excess wastewater to the local waterway. To reduce potential for these discharge events called combined sewer overflows (CSOs), more storage capacity, such as tunnels, must be built. Beyond separate and combined systems, alternatives such as vacuum or septic tanks appear where connecting to a centralized WRRF is too expensive or not appropriate because of high water table, rocky terrain, poor soil conditions, low population density, or other site-related limitations.

Proper sewer main inspections are vital to ensuring conveyance, preventing environmentally damaging overflows of untreated wastewater, preventing structural damage to roads and other infrastructure, and discovering sources of infiltration and inflow. Frequent attributes used to predict or plan maintenance activities for sewer mains include pipe length, pipe material, year of construction, maintenance history, condition assessment history, and slope. Additionally, the components of gravity and pressure pipeline systems are substantially different and consequently require different inspection tools and approaches. The U.S. EPA's publication *National Pollutant Discharge Elimination System (NPDES) Compliance Inspection Manual* provides recommendations on sanitary sewer overflows (SSO) and CSO prevention in Chapters 13 and 12, respectively (U.S. EPA, 2017).

1.3 Lateral Sewer Pipes

Sewer laterals carry wastewater from properties to the public sewer main. Laterals may become clogged and cause basement backups or act as a source of infiltration and inflow, which affects the sewer conveyance capacity and ultimately the WRRF's treatment capacity. As illustrated in Figure 2.2, sewer laterals connecting residences and establishments to public sewers are typically divided into two sections—upper and lower laterals. The upper lateral connects the private property to the cleanout, which is located at the sidewalk or where the property transitions to public property. The lower lateral connects the cleanout to the main sewer line. In some wastewater systems, the entire lateral is maintained by the property owners, while in other systems, property owners are only responsible for maintaining the upper portion, and the municipality maintains the lower portion (Water Environment Federation [WEF], 2016). Property owner responsibility is typically defined in the local jurisdictional ordinance. Examples of important lateral attributes include installation date, location, lateral type (public, private), material, connection type (e.g., factory tap, hammer tap), and diameter.

1.4 Lift Stations

Sewer lift stations, also known as pump or pumping stations, move wastewater to higher elevations using a pressurized system. There are two primary types of lift stations—dry well stations and submersible (wet well) stations.

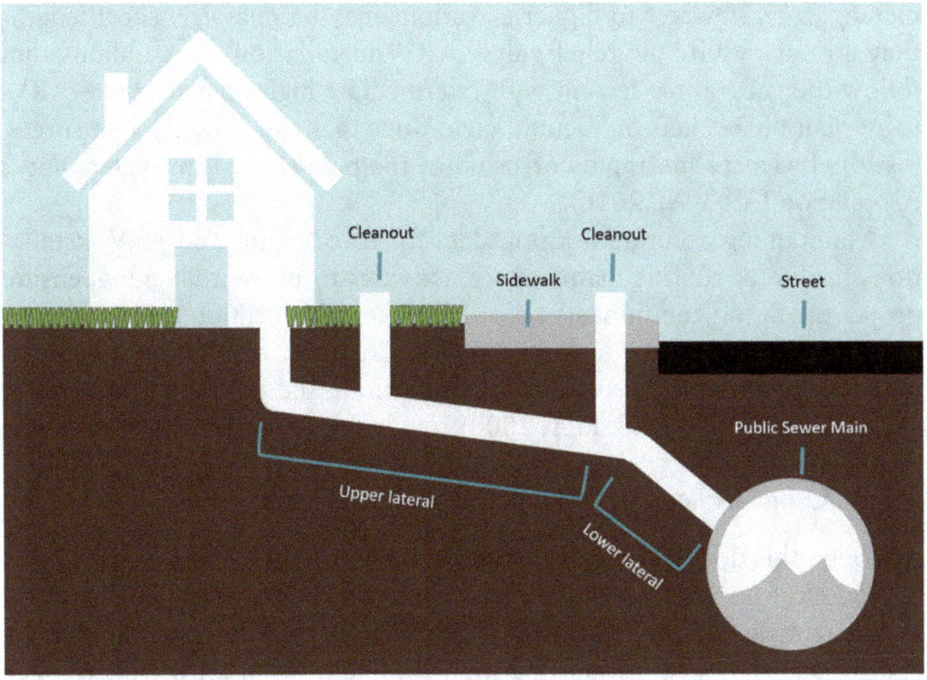

FIGURE 2.2 Upper and Lower Laterals

In a dry well system, an enclosed structure called the dry well stores the pumping equipment and infrastructure. The sewage enters a wet well that stores the wastewater. When the wastewater reaches a predetermined level, pumps lift the wastewater up through pressurized pipes. In a submersible station, also referred to as a wet well station, the sealed pump sits in the wet well directly submerged in wastewater. Although submersible pumps are typically less costly to install and have a smaller footprint, routine inspections become more difficult to perform. A third and less common system are grinder systems, which pulverize the materials in raw wastewater and then convey it to the sewer main. Grinder stations are typically used in problematic locations where foreign materials such as rags and other debris are frequently introduced into the collection system (U.S. EPA, 2002).

1.5 Force Mains

Force mains are pressurized sewer pipelines that convey wastewater from the discharge side of a pump. Pumps or compressors force the wastewater to higher elevation or transport wastewater to the WRRF. Force mains are often useful for reducing the size or depth of gravity sewer systems when reaching depths below ~6 m (20 ft), which can significantly increase cost and complexity of continued gravity sewer construction. Force mains that merely lift wastewater to higher elevations may be relatively short lengths that do not require air relief valve (ARV) installation. Force mains that follow the geographic terrain with intermediate high points require ARVs to prevent air pockets or vacuum conditions that can cause system surges, loss of efficiency, and rapid corrosion of the pipe (American Water Works Association [AWWA], 2016).

Maintaining accurate location data for force mains and ARV installations facilitates routine maintenance activities to prevent clogging, ensure proper operating conditions, and provide accurate location information to avoid potential damage during future construction. In addition, maintaining easement access is necessary for inspections and repairs of all types of conveyance systems (U.S. EPA, 2000).

1.6 Other Collection System Components

Although this document does not allow space for an exhaustive discussion of all collection system assets, several other components beyond the scope of this work include siphons, tide gates, storage facilities, and supervisory control and data acquisition (SCADA)/communication infrastructure. For more details, refer to *Operations and Maintenance of Wastewater Collection Systems* (California State University, Sacramento, 2018).

2.0 PRELIMINARY INSPECTION METHODS AND TECHNOLOGIES

The recent availability of cost-effective preliminary inspection technologies has allowed many wastewater utilities to transition over time from a reactive to a proactive maintenance approach. Rather than using failures in the system or customer complaints as the trigger for maintenance activities, preliminary inspection methods focus on collecting condition assessment information to identify and prioritize where maintenance activities are most necessary. This can reduce maintenance inefficiencies, prevent costly system failures, minimize complaint calls, guide resource deployment, and help determine asset rehabilitation activities. Common types of preliminary inspection methods for various asset types are summarized in Table 2.1.

2.1 Above-Ground Visual Inspections

Where feasible, visually examining collection system assets often acts as an appropriate first step for gathering data on their condition. Visual inspections of wastewater collection systems in this chapter are limited to above-ground

TABLE 2.1 Wastewater Collection System Preliminary Inspection Approach Summary

Asset	Common Problems	Possible Preliminary Inspection Approach
Manholes	Clogs, invert damage, wall damage, offset rings or cover, leaking joints, walls or cover, infiltration and inflow entry	• Visual inspection of covers, walls, steps, and inverts from above • Camera inspection of walls, steps, and inverts from above • Interior visual or camera manned entry inspection • High water level monitors and alarms
Sewer Mains	Clogs, breaks or collapse, joint separation, bottom scour, corrosion, liner damage or failure, cross bores, hits, hydrogen sulfide (H_2S) corrosion, infiltration and inflow entry	• Visual inspection of flows in manholes • Visual inspection of sewer route for signs of depressed or wet ground • Observation of material and debris in flow removed from a pipe blockage or SSO event • Acoustic inspection for blockages (small diameter) • Zoom camera for visual inspection of blockages and general conditions (large diameter) • Observation of contractors or directional drillers working in area • Push camera inspection from manhole or from pipe openings during excavation • Inspect nearby industrial facilities and food service establishments • Flow monitoring • Strategic surface water sampling along stream segments
Lateral Sewer Pipes	Clogs, breaks or collapse, joint separation, corrosion, cross bores, hits, H_2S corrosion, infiltration and inflow entry	• Visual inspection of flows in cleanouts • Visual inspection of lateral route for signs of depressed or wet ground • Observation of contractors or directional drillers working in area • Flow monitoring and smoke testing • Review of local food service establishments • Review of permitted industrial users

(continued)

TABLE 2.1 (*Continued*)

Asset	Common Problems	Possible Preliminary Inspection Approach
Lift Stations (Pumping Stations)	Pumps, controls, or alarms mechanical or electrical failures	• Routine operator visual inspections • Automatic monitoring and alarms • Pump run time monitoring and analysis • Periodic infrared camera monitoring • Flow monitoring • Physical inspection (e.g., H_2S corrosion)
Siphons	Grit settlement, infiltration and inflow entry	• Visual inspection of entry and exit points • Observation of material and debris in flow during a cleaning operation • Dye testing (if siphon goes under a waterway) • Flow monitoring
Diversion Structure	Corrosion, equipment failure (if automated), debris damage	• Visual inspection of flows from hatch openings • Interior visual manned entry inspection, if possible, with or without camera inspection • High water level monitors and alarms • Flow monitors to ensure proper flow diversions
Lateral Connections	Leaks, offset connections, protruding connections	• Visual push camera inspection during excavation • Lateral cameral inspection during a closed-circuit television (CCTV) inspection of the sewer main
Cleanouts	Open cleanouts, clogs, breaks, hits, infiltration and inflow entry	• Visual or camera inspection from above • Lateral route monitoring for missing cleanout caps • Smoke testing
Force Mains	Joint separation, scouring from sediment and sand, collapse, life cycle (age)	• Physically walk the force main route for signs of leaks or uneven settlement • Flow monitoring to assess flow balance and trends • Force main assessment through internal inspection technologies

facilities. Chapter 3 of this publication discusses technologies used to prevent wet weather overflows and technologies used for overflow response, and Chapter 4 of this publication goes into greater detail for below-ground inspection techniques.

2.1.1 Manhole Visual Inspection

Routine visual manhole inspections help record current asset conditions as well as identify potential infiltration and inflow sources, cleaning needs, or structural deterioration. These inspections also help to verify inventory details. Operators systematically observe the general condition of the various manhole components. The condition is then consistently documented for maintenance needs observed, such as issues regarding hydraulics such as surcharges, flow level, or flow state. Manhole inspections are often best conducted when the ground is saturated to examine deficiencies that may not be observable during dry weather. For example, leaky covers, leaks from broken joints, or leaks from cracks may not be noticeable in dry weather conditions, whereas in wet weather, water flow will be visible. Reference Chapter 5 of this publication for more information on manhole rehabilitation and replacement.

2.1.2 Lift Station Visual Inspection

Lift stations require periodic checks to ensure reliable operation and performance. Some maintenance activities, such as visually inspecting for sludge and grease accumulation, alarm system operation, and indications of high-water levels should be conducted frequently. Regularly inspecting pumps and removing grit are also important for ensuring performance. In addition to visual inspections, maintaining a recorded log of pump running time helps with analyzing pump wear and tear, detecting inconsistent pump down times that may indicate system problems, and optimizing energy usage. If confined space entry is necessary for inspecting the lift station or a manhole, all confined space entry requirements and regulations should be followed, per guidelines such as *Occupational Safety and Health Standards, Permit-Required Confined Spaces* (U.S. Department of Labor, 2011).

2.2 Preliminary Inspection Technologies

Operations and maintenance personnel frequently open manholes as part of routine flow control activities. Therefore, it is useful to equip field crews with standard manhole inspection forms or access to an electronic collection device to record conditions that may require additional maintenance activities.

2.2.1 Manhole Three-Dimensional Scanning

When gathering more detailed information is warranted, manhole scanning equipment provides more detailed information. Automated three-dimensional manhole scanning devices utilize several cameras and a laser to scan and visually depict manhole wall conditions. The scanning equipment is deployed from a tripod situated on top of the manhole opening and the inspection data is typically collected within minutes. Three-dimensional scans provide detailed quantitative data for rehabilitation analysis, hydrological surveys, and general condition assessment.

2.2.2 Zoom Cameras

Zoom cameras, or pole cameras, provide video or still images via a pole lowered into the manhole. Operators can observe manhole wall conditions and the pipe conditions from the perspective at the bottom of the manhole. Zoom cameras provide a quick way to screen. However, because zoom cameras are highly dependent on sight distance, obstructions such as water vapor, debris, pipe misalignment, or pipe curves limit their effectiveness in smaller diameter pipes. Because visibility is reduced as pipe diameter decreases, application of zoom cameras should consider diameter size and recognize performance improves with larger diameter sewer pipes.

2.2.3 Acoustic Inspection Technology

Acoustic technology uses two components that sit on top of adjacent manholes. One device generates a series of audible tones, while the device on the other end receives and interprets the sound signal to generate a blockage assessment. Acoustic inspection provides a quick and cost-effective screening tool for small diameter gravity pipes 150 to 460 mm (6 to 18 in.).

3.0 DEPLOYABLE MONITORING TECHNOLOGIES

Monitoring sewer collection systems flow facilitates the evaluation of the systems conveyance performance under wet weather conditions, pinpoints sewer capacity limitations, serves as the data inputs for hydraulic models, and drives prioritization for follow-up inspections, maintenance, and repair of sewer assets. Selecting the appropriate monitoring equipment for each application requires an understanding of the intended objective and the technologies available to achieve that end (Collection Systems Committee, 2019).

The appropriate application, each location's unique characteristics, and environmental factors can all affect the accuracy of monitoring data gathered. For example, grease, foam, sediment, debris, turbulence, and vapors in the air space can all affect sensor readings differently. Software is needed for monitoring equipment setup and for post-data processing, visualization, assessment, and creating reports. Modern flow monitoring equipment is typically wireless, making for easier installation and remote sensing. Integrating a SCADA system can help to automate data collection and enable real-time monitoring capabilities. Additionally, creating a quality control/quality assurance program integrating proper calibration and equipment maintenance is critical for ensuring data quality. More information on post-data visualization tools is covered in Section 3.1.5, Flow Data and Scatter Plots.

3.1 Open Channel Monitoring Technologies

Open channel flow monitoring technologies include fixed hydraulic structures such as flumes, weirs, and area velocity flow meters (A/V meters). A general overview of each technology is provided in this section, and Table 2.2 should be referenced for comparison and considerations.

3.1.1 Flumes and Weirs

When flumes are combined with a secondary level sensor or staff gauge, they can establish flowrate and a subsequent volume over time. Although there are many types of flumes, their general functionality relies on concentrating fluid in an open channel through a fixed structure with known discharge tables by narrowing sidewalls at the flume throat. The resulting level of fluid in the flume can be correlated to a known flowrate derived from discharge tables associated with each flume or weir. Consideration must be given to the upstream fluid characteristics to ensure appropriate hydraulic conditions exist that are favorable to obtaining representative data.

3.1.2 Area Velocity Flow Meters

Area velocity flow meters are an integrated multisensor configuration designed for use in standalone remote applications, treatment facilities, and process control applications. Area velocity meters calculate flow in open channels by using pipe size in tandem with the integrated sensors measuring depth of flow (D) and velocity (V) of the fluid. The data inputs from the velocity and level sensors, combined with pipe size, establish the wetted surface area (A) enabling the final calculation for flow or quantity (Q). In summary, Area multiplied by Velocity = Quantity or simply, $AV = Q$.

The velocity sensor uses the Doppler effect to determine the speed of the fluid through the pipe. This is done by transmitting a known frequency

TABLE 2.2 Comparison and Overview of Open Channel Monitoring Technologies

	Flumes, Weirs, & Hydraulic Structures	Area Velocity Flow Meter (A/V Meters)	Ultrasonic Level Sensor (ULS Sensor)
		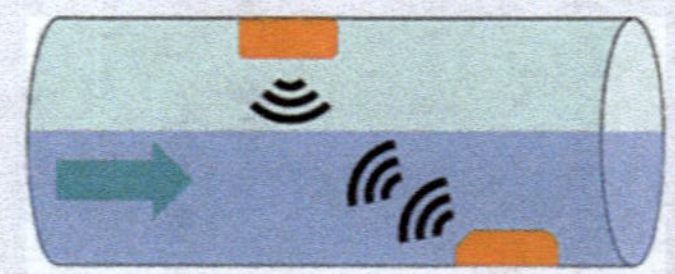	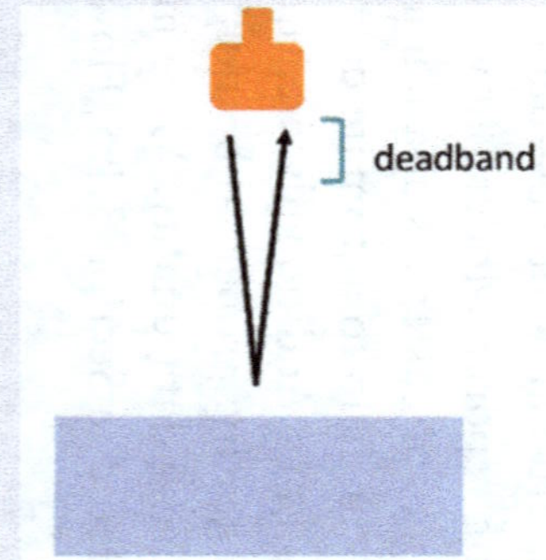
Description	Fixed hydraulic structure that requires level inputs from a secondary sensor to calculate and establish volume based on a discharge table.	Establishes flow by obtaining velocity, depth, and pipe size data to calculate flow. Types include contact or in-flow (submerged) or non-contact.	ULSs emit and receive high frequency sound waves above the fluid. Fluid level is determined by measuring the delta between frequency transmit time and return echo.
Application & Benefits	• Often used on larger pipes • Easy to maintain post construction • Excellent data repeatability	• Easily deployed in existing sewers • Can measure reverse flow conditions	• Generally small, inexpensive, lightweight, easy to install and maintain • Mature technology • Can be used to assess the site hydraulic conditions to mobilize O&M crews for cleaning

Limitations	• Cannot measure reverse flow • Upfront cost can be high (construction) • Requires a secondary measuring device for level • Pipe slope and hydraulic conditions affect accuracy and overall suitability	• May require confined space entry • May not measure velocity on extremely low flows • In-flow type can be problematic in a larger diameter pipe • High velocities or debris can rip out sensors • Poor hydraulic conditions affect accuracy • Contact A/V meters can cause underreporting of velocity in larger diameter pipe because of limitations in range, which can inhibit ability to measure where velocity is greatest	• Flow conditions such as turbulence, foam, or floating debris can cause inaccuracies • Limited by dead band (unable to measure first few inches from sensor)

into the fluid and capturing the return echo off the suspended particles and bubbles in the fluid and calculating the shift in frequency from the original known transmission. The frequency shift between transmitted and received correlates to the velocity or speed of the fluid through the section of pipe.

Area velocity meters fall into two categories: submerged or in-flow and non-contact style. Non-contact styles are also used in applications unsuitable for staff to enter. Non-contact style monitors are positioned above the flow, mounted to the top of the pipe or in the manhole riser, and focused downward to the fluid. Depths are generally determined by an ultrasonic sensor until the pipe becomes surcharged to the top of the pipe. Above the surcharged levels, a redundant depth sensor can be used to measure depth of flow. Velocity is determined by either radar sensor, or a laser sensor focused on the surface or attenuated just beneath the surface of the fluid.

3.1.3 Ultrasonic Level Sensors and Submerged Probe Level Sensors

Level sensors are used to measure depth in open channel sewers and lift station wet wells and can help determine when crews should be mobilized for cleaning. They can also be used to calculate flow estimates using the Manning Formula or by correlating current reported level data with historical reported flow/level data from the same location. Sewer level data captured during rain events can also be useful for understanding a sewer system's wet weather response by comparing dry weather and wet weather levels. Unique site conditions and hydraulics should be given consideration for each application and sensor configuration.

Ultrasonic level sensors (ULSs) are installed above the fluid surface. They emit and receive high frequency sound waves. Fluid level is determined by measuring the delta between the transmit time and return echo from the surface. Because ultrasound reflects off surfaces, site conditions such as turbulence, steam, foam, and floating debris can affect accuracy. For accurate measurement, ULSs require an unobstructed "line of sight" to the flow. Ultrasonic level sensors cannot measure within the first few inches of their transmitter, which is called the dead band. Should water levels rise and submerge the ULS, it can no longer detect the level in which case a backup or surrogate level sensor should be given consideration.

Pressure transducer sensors are submerged under water and measure the pressure at the bottom of the pipe, tank, or well to determine the fluid level. Pressure transducers can be affected by sediment and are vulnerable to chemical discharges and may require confined space entry. Pressure transducer sensors are a suitable level sensor option when site and hydraulic conditions are turbulent or when steam, foam, and/or floating debris are present. Pressure transducers can be used in tandem with ULSs as a redundant level input.

Manhole (depth-only) sensors capture sewer levels. These are systems that include a level sensor, power source, control board, and cellular modem. One type of system incorporates floats designed to provide alarms for surcharge event warnings on potential SSOs. These are low-cost systems for the purpose of proactive alarming on manholes in which they are deployed.

3.1.4 Micro-metering

Micro-metering is a monitoring deployment strategy used for pinpointing and evaluating the effects of rainfall-derived infiltration and inflow (RDII) into the sewer conveyance system. Micro-metering focuses on small sections of a sewershed basin, perhaps a few thousand feet of pipe based on the Pareto Principle. This method can make use of using flow or level measuring devices that are deployed for a short duration of time to identify rapid flow rise or drop, which indicate potential sources of RDII.

3.1.5 Flow Data and Scatter Plots

Time-series data is typically stored in a database that is accessible via data visualization software. Depending on the software and data-throughput configuration, data can be viewed real-time for SSO prevention, tracking, and reporting or for post-deployment analysis. Monitoring data software visually displays data in hydrographs, tables, or scatter plots. Scatter plots visually depict velocity and depth measurements on an x-y axis. The resulting pattern can be a useful visual tool in the office for diagnosing problematic hydraulic conditions.

3.2 Flow Monitoring—Closed Conduit and Pressurized Pipes

Closed conduit flow meters use similar mechanisms and fundamental calculations as the open channel flow meters to establish flows and provide data inputs for system assessment. They can help reduce SSOs by validating pump curves and alerting operators of potential system failures. A general overview of each technology is discussed in this section with Table 2.3 comparing applications and considerations for flow meter types.

3.2.1 Insertable Doppler

Insertable doppler flow meters are an accurate, low maintenance, and cost-effective option designed for use in closed conduits. These meters can be installed on existing piping using the hot tap method with minimal space

TABLE 2.3 Comparison and Overview of Closed Conduit Monitoring Technologies

	Insertable Doppler (Doppler Ultrasonic Insert Velocity Meter)	Transit-Time Flow Meter (Clamp-on or Strap-On Flow Meter)	Magnetic Flow Meters (Mag Meter)
	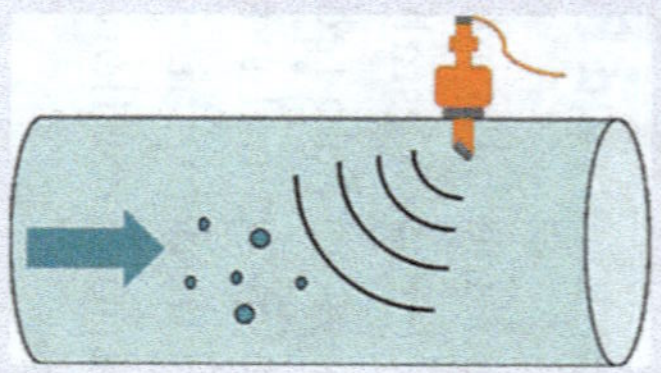	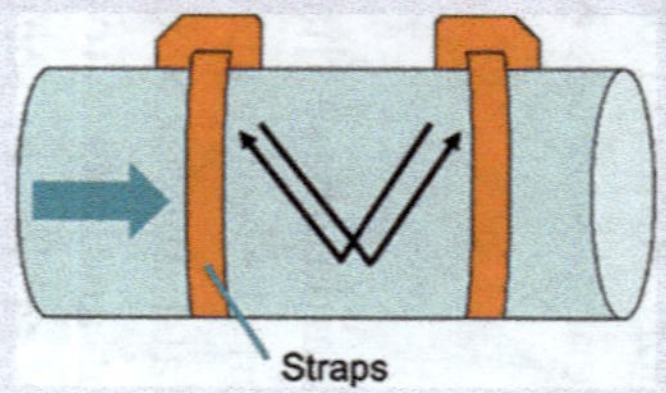	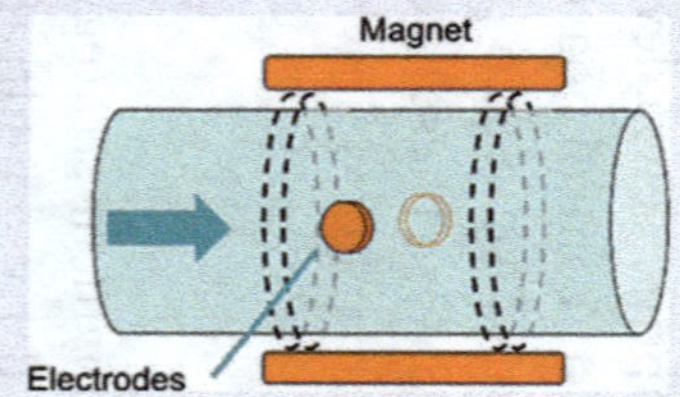
Description	Establishes volume by emitting a frequency into the fluid and measuring the return Doppler shift off particles; the shift is proportional to fluid velocity.	Establishes volume by transmitting sound waves through the pipe wall and the transmit-time of the sound wave is proportional to the fluid's velocity.	Establishes volume by generating a magnetic field and measuring the voltage change resulting from a conductive fluid passing through the magnetic field.

Application and Benefits	• Commonly selected for use on larger pipes • Cost not affected by pipe size • Can be installed on existing pressurized pipes using hot tap method	• Easily installed and replaced • Non-intrusive; clamps on the outside of the existing pipe • Different sensor configurations (Z-path, W-path, V-path) • Can be used in temporary and remote standalone applications	• Good for applications requiring precision and accuracy • Good data repeatability
Considerations and Limitations	• Sensors contact the fluid • Requires penetrating the pipe	• Pipe bends and connections before and after can affect signal returns and accuracy • Signal quality may not be effective on some pipe materials or pipes with heavy tuberculation or deposits	• Pipe bends and connections immediately before and after can affect post-data accuracy • Nonconductive fluid may be problematic • Cost of meter tends to go up in scale with size of the pipe • Difficult to calibrate/validate

requirements. Insertable doppler flow meters function similarly to the A/V meter technology discussed in Section 3.1.2 in that they use the Doppler effect to determine fluid velocity.

3.2.2 Strap-On Ultrasonic Flow Meters (Transit-Time Flow Meters)

A transit-time flow meter also referred to as strap-on or clamp-on ultrasonic flow meter, provides a cost-effective option for measuring flow in a closed conduit. The transit-time system is composed of two attachable transducers that clamp to the exterior of the pipe. Ultrasonic waves are transmitted through the pipe wall both with and against the direction of flow, and the sound waves transit-time to the receiving sensor can be correlated to the fluid velocity.

3.2.3 Electromagnetic Flow Meters

An electromagnetic flow meter, also called mag meter, magnetic, or induction flow meter provides a low maintenance, accurate technology for measuring flow in a closed conduit. The unit contains two magnetic coils located on opposite sides of the conduit, which generates a magnetic field. Two opposing sensing electrodes used to measure voltage are located perpendicular to the magnetic coils. When a conductive liquid travels through the magnetic field, it generates a voltage, which is measured by the sensing electrodes. This voltage reading is proportional to the velocity in the pipe.

3.3 Rain Gauges

Rain gauges provide precipitation intensity data used to evaluate conveyance system performance under various wet weather conditions. Rainfall data can be used to identify RDII, which serves as an indicator of collection system maintenance issues, as a process control input, and as a metric for regulatory reporting. Collecting representative precipitation data is an important element for understanding how weather affects the collection system performance and hydraulic conveyance capacity. Rain gauge data is also commonly used for NPDES reports and CSO facilities and establishing storm magnitude after a rain event. Free rainfall data is available to the public from federal government agencies like the U.S. Geological Survey (USGS) and National Oceanic Atmospheric Administration. Rain threshold notification alarms can be set up online at no cost to the end user. However, USGS rain gauges are dispersed over a wide area and may not yield precipitation resolution needed on localized storms and microbursts that occur in a small geographic area.

Rain gauges primarily fall into two primary categories: non-recording or self-recording. Non-recording rain gauges have a funnel that feeds into a receiving cylinder, and the water level is manually recorded by visual observation. During heavy rainfall, non-recording rain gauges can be disadvantageous because they may require multiple manual readings a day to ensure that the receiving cylinder does not overflow and cannot calculate time-based rainfall intensity. Self-recording rain gauges include tipping bucket, weighing bucket, or floating types. Figure 2.3 illustrates a tipping bucket rain gauge.

To obtain representative rainfall data, rain gauge site locations should consider wind exposure and overhead obstructions. For proper installation, the gauge should be placed on a level surface that is not subject to excessive wind disruption. With respect to rain data validity and rain gauge site density over a geographic area, all rain gauges have limitations and shortcomings that should be taken under advisement. Rain gauges only represent point estimates of rainfall, and they can become clogged, become covered by surrounding vegetation, or lose power.

Gauge adjusted radar rainfall (GARR) may add additional value when assessing conveyance system performance under wet conditions. GARR amalgamates data inputs from local radar and rain gauges. Having multiple

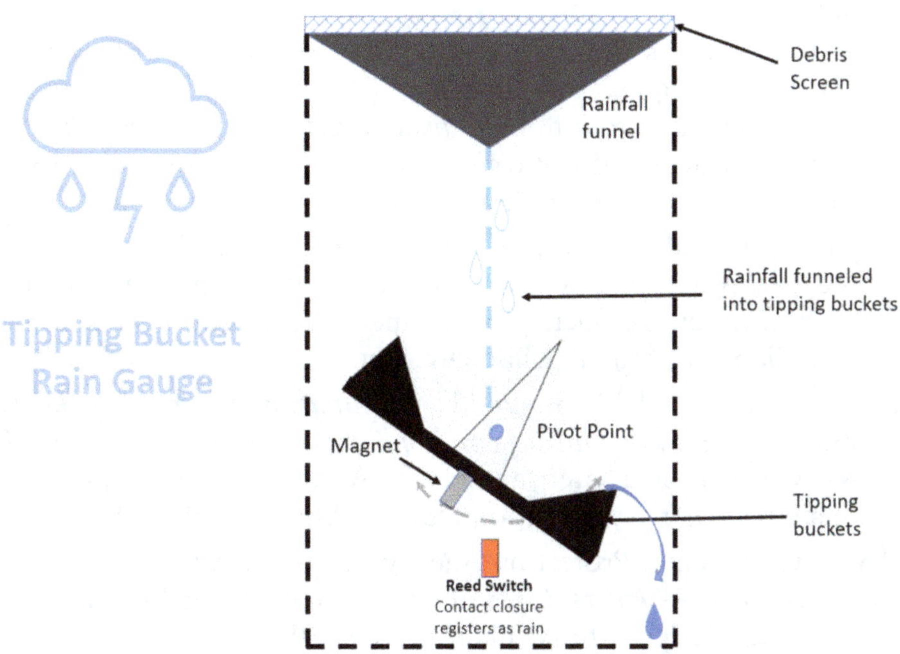

FIGURE 2.3 Tipping Bucket Rain Gauge

data input sources helps fill gaps where no gauges exist, gives a clearer idea of localized intensity, corrects, and validates both data inputs and can provide a more accurate representation of a rain event for a hydraulic model.

3.4 Monitoring Equipment Calibrations

During the commissioning phase of deploying monitoring equipment such as flow meters, rain gauges, and level sensors, a calibration should be conducted to ensure the instrument functions properly and accurately reflects the current site conditions. In addition, reported readings should be confirmed to be within an acceptable range of deviation. Calibration records should be maintained to evaluate trends and serve as historical reference for data validity and identifying possible equipment problems.

4.0 REFERENCES

American Water Works Association. (2016). *Air-valves: Air-release, air/vacuum and combination* (2nd ed., Manual of Water Supply Practices M51).

California State University, Sacramento. (2018). *Operations and maintenance of wastewater collection systems* (8th ed.). Office of Water Programs.

Collection Systems Committee. (2019). *Fact sheet—Sanitary sewer flow monitoring and data analytics* (WSEC-2019-FS-011). Water Environment Federation. https://www.wef.org/globalassets/assets-wef/direct-download-library/public/03---resources/wsec-2019-fs-011---csc---flow-monitoring-and-data-analytics---final.pdf

Hughes, J. B. (Ed.). (2009). *Manhole inspection and rehabilitation* (2nd ed., ASCE Manuals and Reports on Engineering Practice No. 92). American Society of Engineers, Committee on Manhole Rehabilitation of the Pipeline Division.

U.S. Department of Labor. (2011). *Occupational safety and health standards, permit-required confined spaces* (Standard No. 1910-146). Occupational Safety and Health Administration. https://www.osha.gov/laws-regs/regulations/standardnumber/1910/1910.146

U.S. Environmental Protection Agency. (2000). *Wastewater technology fact sheet—Sewers, force main*. Office of Water. https://www3.epa.gov/npdes/pubs/force_main_sewers.pdf

U.S. Environmental Protection Agency. (2002). *Wastewater technology fact sheet—Sewers, pressure*. Office of Water. https://www.epa.gov/sites/production/files/2015-06/documents/presewer.pdf

U.S. Environmental Protection Agency. (2017). *NPDES compliance inspection manual* (305-K-17-001). Office of Enforcement and Compliance Assurance. https://www.epa.gov/sites/production/files/2017-01/documents/npdesinspect.pdf

U.S. Environmental Protection Agency Region 1. (2017). *Developing a manhole or catch basin numbering system.* https://www3.epa.gov/region1/sso/manhole-id.html

Water Environment Federation. (2016). *Private property infiltration and inflow control.*

5.0 SUGGESTED READINGS

U.S. Department of the Interior, Bureau of Reclamation. (2001). *Water measurement manual—A water resources technical publication* (revised reprint). https://www.usbr.gov/tsc/techreferences/mands/wmm/

3

Collection System Operation and Maintenance-Related Technologies

Charles H. Starling III, PE, BCEE; Stacy N. Hayden, PE; & Brock A. Jones, PE

1.0 SEWER STOPPAGES AND OVERFLOWS

Most collection system operators face a continuous battle with sanitary and combined sewer overflows (SSOs and CSOs). This chapter focuses on available tools and technologies used in the collection of system performance data;

development of an accurate system database; recommended information to collect; methods for analyzing the collected data; the use of hydraulic models for predictive and preventive measures; and effective response preparation to address SSOs and wet weather CSOs when they occur.

1.1 Frequency, Location, and Monitoring

A critical component of understanding the performance of sewer collection and conveyance systems begins with access to historical performance data. Individuals are often aware of recurring issues within a collection and conveyance system; however, this information is often not recorded in a central database where it is easily accessible to others within an organization. Being proactive in record keeping will save time and frustration and provide for a more effective means for proper analysis of system operation and performance.

Tools used for real-time monitoring can include pressure transmitters, flow monitors, flow meters, and level monitors. Real-time monitoring and reporting tool technologies are available and are becoming less expensive to purchase and easier to operate and maintain. Alarm settings on monitoring equipment can be adjusted to allow operators to receive warnings to improve response times and to obtain the appropriate balance between critical and noncritical notifications.

1.2 Reporting to Regulators and the Public

The primary areas of focus in reporting to regulators and the public differs. In reporting to the public, the primary focus is on safety and alerting the public in a timely manner. The public should be aware of the location of an SSO and how it could affect their activities and potential safety. Signage, caution tape, or barriers should be installed while the emergency response and cleanup are occurring to prevent contact. Portable message boards may be needed to warn the public to avoid entering or playing in waterbodies or pools of water potentially contaminated with wastewater.

A delay in reporting SSOs to the public can result in increased safety risks and diminish public confidence in the utility. Tools used for public notification include utilization of mass notification systems, which may include email, social media, reverse 911, and news outlets. Each means of notification will improve the level of detail and timing of the reporting process. In addition, consideration should be given to match the notification response with the severity of an SSO. For example, you would respond differently to a grease choke in a cul-de-sac in a residential neighborhood versus a large diameter force main break.

Reporting to the regulators is mandated by state laws and regulations that focus not only on public safety, but also on mitigating the effects of the SSO, identifying of the SSO cause(s), and taking corrective actions to reduce or eliminate future SSOs from occurring. To best facilitate timely reporting, consideration should be made to develop standardized reporting mechanisms to consistently provide accurate and complete reports. Without standardization, reporting can be time consuming, is often insufficient, and may be less descriptive than required. Standardization of reporting methods can be as simple as a paper-based template with critical items to note per occurrence to a more comprehensive field accessible geographic information system (GIS) application.

1.3 Public Versus Private

One of the most difficult issues to address in collection systems is dealing with privately owned assets such as the private side of the lateral or sewers and lift stations that were never deeded to the utility or municipality. Although public side infrastructure is more easily accessible and can be maintained with a well-structured program, the public side is directly affected by the performance of the private side utilities. For instance, a well-maintained gravity sewer system can suffer from peak flows from rainfall-derived inflow and infiltration (RDII) simply from failed private side laterals. Few utilities have initiated a private side lateral program, or if so, are in the early stages of addressing private side issues both monetarily and legally. Utilizing data collected through a sewer system evaluation survey program, coupled with flow monitoring response times, can help identify private side issues contributing to peak flow conditions and the corresponding SSOs. Studies such as *Reducing Peak Rainfall-Derived Infiltration/Inflow Rates—Case Studies and Protocols* (Water Research Foundation, 2003) have shown peak wet weather SSOs are frequently directly related to the private side laterals.

In addition to private lateral issues, many utilities receive wastewater from private sewers or lift stations built to serve subdivisions, multifamily complexes, commercial shopping centers, or industrial developments where ownership of the facilities was not deeded to the utility upon completion. Operation and maintenance of such systems remains the responsibility of the property owner's association or the developer who are frequently ill-prepared and poorly funded to assume such responsibility.

1.4 Tracking Causes

As mentioned, a key aspect of a well-maintained collection system is directly related to record keeping. Most operators are aware of problem areas within their collection system; however, when conducting a root cause analysis, critical data is often unavailable. For example, data on a SSO is sometimes limited to a manhole located near a certain address or street intersection with a note that an overflow occurred with an estimated discharge. Supporting information in this instance should include a manhole identification number, location in a GIS, photographs of the SSO, information regarding the preceding or weather conditions at time of occurrence, number of previous SSOs, date of occurrence, cause of occurrence, and so on. Frequently, the actual cause of the SSO may not be at the overflow or discharge location.

Data collected during these events is necessary to help determine the root cause and to help develop corrective actions necessary to reduce or eliminate recurring sanitary sewer discharges. Inaccurate or incomplete data collection can equally serve to reduce corrective action success and prevention of future SSOs. A vital component needed to properly implement this process is training of the individuals and conveying the importance of accurate and complete data to those responsible for data collection.

The simplest form of tracking SSOs is through regulatory reporting forms that are often submitted to the agency and then placed into a file. Although this may meet compliance, it does not provide a means to efficiently access information necessary to perform a system evaluation. Lack of a central and accessible data repository often results in key system operational knowledge that is not fully shared with others within a utility. Recording complete and accurate data can be achieved by providing access to the GIS database–linked applications that have standardized reporting forms to prompt the observer for data collection. Having this access will help to ensure that recording targeted information is consistently collected during an emergency response and available to others within a utility for post-event analysis.

Development of a GIS database should not be viewed as an overwhelming or cost-prohibitive venture. The key to initiating a GIS program is to begin with a basic layout of the existing primary components of collection and conveyance system. The initial system could be limited to approximate locations of large diameter force mains, large diameter gravity mains, manholes, and limited pipe attributes. Once a base layout of the system is developed, routine updates can be made that include corrections to alignment, inclusion of smaller diameter mains, updates in pipe diameters, materials, manhole numbering schemes, and pipe inverts. This should be considered a continuous process where updates can be performed upon data availability. Consideration should be made to implement quality control measures for updating the GIS database. The quality control measures should be sufficient to help improve the accuracy of information collected, but not too restrictive to deter updates. Development of workflow procedures at the onset of GIS development can help to ensure consistency and accuracy of the data.

Creation of attribute tables to include performance data, such as SSO occurrences, causes, cause/asset locations, dates, photographs, and so on, is available. As the database is developed and relevant data is added to the system, GIS tools can be used to query data, whether it is asset data in terms of system properties or querying the number of SSOs within the system. Beginning with the base use of a GIS sewer system database and transitioning to a more detailed and accurate database, will provide information necessary for moving from a reactive approach to system maintenance to the predictive and preventive mindset.

There are GIS tools available, as illustrated in Figure 3.1, that allow for closed-circuit television (CCTV) sewer inspection data to be linked to a GIS database. Linked data can include video, pipeline assessment certification program coding, and grading assigned to a respective sewer asset. Field data can also be linked to GIS through mobile data collection applications using smartphones or tablets. This integration process allows the operator to easily view record information, both data and video, on a specific asset.

An example of a query to address repeat or chronic SSOs using a GIS database might include identifying all manholes with more than two SSOs

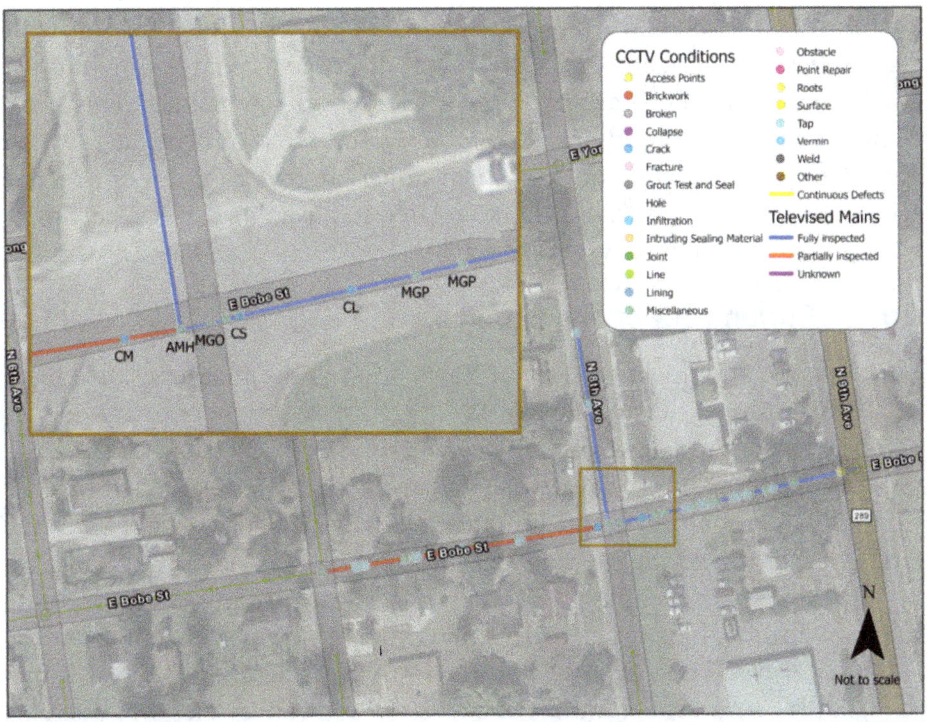

FIGURE 3.1 GIS-Based CCTV Tracking Example

within a 12-month period. Depending upon the size of the system, available data, or local regulatory requirements, the specific query needs will vary. However, the query can be adjusted to obtain the range of data that is of interest. Another query might include listing all manholes where SSOs occurred with a recorded rain event exceeding 1 inch over a 24-hour period. Successful use of a GIS database is directly dependent upon the available data included within the database and its accuracy.

Establishment of a GIS database also allows utilities to track backup complaints, blockages, "hot spot" areas, or odor complaints that may result in an SSO event. Such tracking can help prevent SSOs as well as facilitate communications between operators, engineers, and managers. To be effective, policies requiring digital as-built submittals from contractors and field asset correction submittals should be implemented so the GIS databases remain updated with correct location and asset data.

2.0 PREVENTING DRY WEATHER STOPPAGES AND OVERFLOWS

Identification and tracking of dry weather overflows can prove to be an effective predictive tool for identifying common maintenance-related SSOs. Dry weather overflows are most often related to blockages because of structural issues; fats, oils, and grease (FOG); roots; or an accumulation of other solids. Issues with FOG are normally attributed to commercial areas, multifamily residential (apartments, condominiums, etc.), or sewer mains that have defects such as a sag or insufficient slope. Another contributor to sewer blockages is "flushable wipes." Issues related to flushable wipes is becoming a frequent topic among system operators.

2.1 Tracking Sewer Flow Conditions

Identification and tracking of sewer flow conditions is another critical piece of information. As the availability of data increases, more effective queries can be executed to identify manholes where SSOs occur. Once these areas are identified, the duration between occurrences can be estimated or continuously monitored, and appropriate actions can be taken to minimize the likelihood of future SSOs.

Tracking of dry weather SSOs can also help identify areas to incorporate additional technologies such as level monitors, flow monitors, and acoustic sensing devices. These technologies can provide operators with real-time data on the operating conditions of collection and conveyance systems. This may be helpful in areas of chronic SSOs to determine maintenance intervals and alarm set points that can alert operators of irregular conditions. Implementing these technologies may improve response time to address potential SSOs as well as provide additional data on the system performance.

2.2 Source Prevention for Dry Weather Stoppages and Overflows

Dry weather stoppages are often a preventable SSO; however, prevention is more easily achieved through accurate tracking of readily accessible historical data. Without retaining the history of system performance, locating the problem areas will be a difficult process and would rely on the memory and communication of the system operator(s). Development of an accurate GIS database should include attribute data, historical SSOs with corresponding weather conditions, dates and causes of SSOs, and locations of occurrences. An accurate database provides anyone with general GIS database skills with the information to easily identify dry weather SSO locations and frequency of occurrence. Identification of patterns can be accomplished regardless of familiarity with the collection system and thus, use this knowledge to develop a maintenance and monitoring plan to prevent future overflows. Such tracking systems should be updated with more accurate causes of SSO events following completion of the root cause analysis to ensure proper preventive measures are identified and implemented.

Recording the cause(s) of dry weather overflows (roots, flushable wipes, FOG, etc.) can help improve the focus of public outreach programs. Using this data, a density of occurrences can be developed (sometimes referred to as "hot spot" analysis) in GIS. This can be performed to help identify customers to focus communication efforts. Effective communication to the customers can be accomplished through flyers, notices included with sewer bills, publication on a utility website, or by other means. Figure 3.2 shows how areas of the sewer system vary in SSO density.

3.0 PREVENTING WET WEATHER OVERFLOWS

Wet weather overflows are directly related to system performance during a wet weather event. Understanding the parameters that result in a wet weather overflow will provide the information necessary to address the limitations of the collection system. These parameters include defining the minimum rain event that results in a wet weather overflow at a respective location. Available methods and tools that are helpful in defining these parameters are identified in the following.

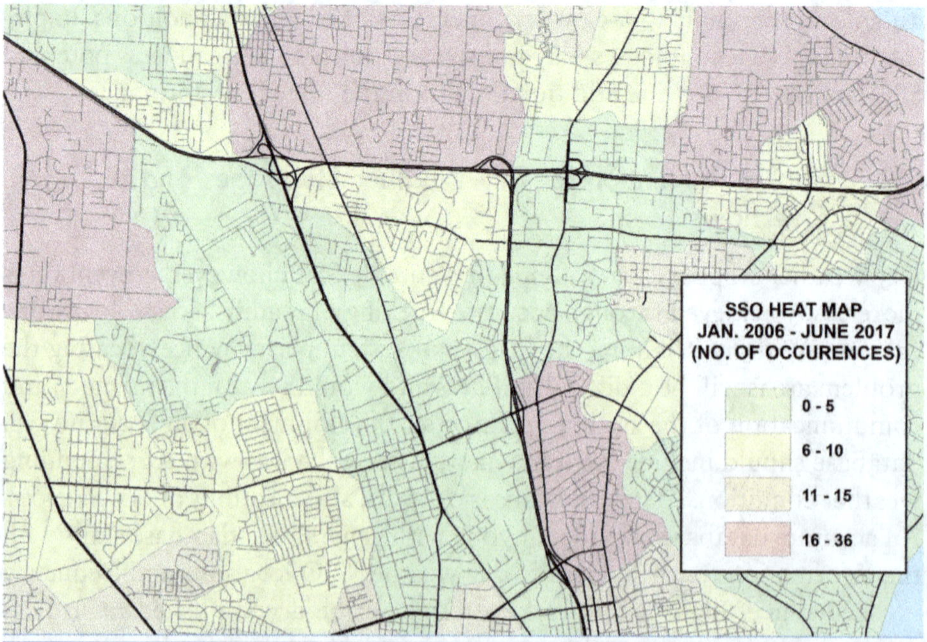

FIGURE 3.2 SSO Heat Map Example

3.1 Hydraulic Model Technologies

Hydraulic modeling technology has advanced over the years and can be incorporated into daily operating scenarios. The use of hydraulic models can vary upon actual needs and can serve as a simple steady-state tool for evaluating peak flows, to complex extended period RDII dynamic models simulating a defined storm event. Additionally, an extended period model can be developed to use integrated real-time weather forecasts to simulate a collection system response to a forecasted storm event.

Although models are useful tools for the management of collection systems, experienced professionals recognize that hydraulic models are often limited by inherent assumptions and conceptual representations of complex processes. The selection of an appropriate hydraulic modeling software is key to the success of any program that seeks to apply models for planning assessments, design, or operational improvements. For example, more advanced hydrologic and hydraulic modeling capabilities are necessary to support an analysis where storage and flow routing effects are important and system performance for various event conditions is to be defined with a greater level of accuracy. Collection system models should be coordinated with water resources research center models for facility headworks or other facility capacity limitations and adverse effects.

Within these more advanced modeling methods, various techniques to better represent system performance under multiple-event conditions and over long-term simulations continue to evolve. One limitation is that extrapolations to larger storms than those observed in the model calibration monitoring period or to portions of the sewer system upstream of the flow monitoring locations will have greater uncertainty that should be recognized in the interpretation of model results.

Another limitation of sewer system models is the risk of software or data not being regularly updated. When investing in a hydraulic model tool, considerations should be made to maintain and update the model regularly. A long-term program should include periodic reviews of model performance and staff training needs, and as required, update the model to ensure its continued value as an effective tool within the program. Figure 3.3 provides a general workflow to follow when implementing a hydraulic model.

3.1.1 Recommended Software Features

Hydraulic model software requirements will vary depending upon application. Key items to consider include type of analyses to be performed and level of complexity of the analysis. Complexity ranges from steady-state analysis (snapshot analysis) to a complex real-time hydraulic modeling over a defined time period. Increasing the number of central processing units (CPU), independent graphics cards, and using sufficient random-access memory (RAM) will help to improve the performance of complex hydraulic model simulations. However, it is important to understand the limitations of the hydraulic model used and if these options are available.

3.1.2 Continuous Versus Event-Based Calibration

Proper calibration is necessary for providing a reliable hydraulic model. Event-based calibration can limit the model accuracy to specific conditions and can result in reduced accuracy during other event simulations.

Utilization of flow and rainfall monitoring data for RDII hydraulic model calibration is necessary to fully understand the collection system

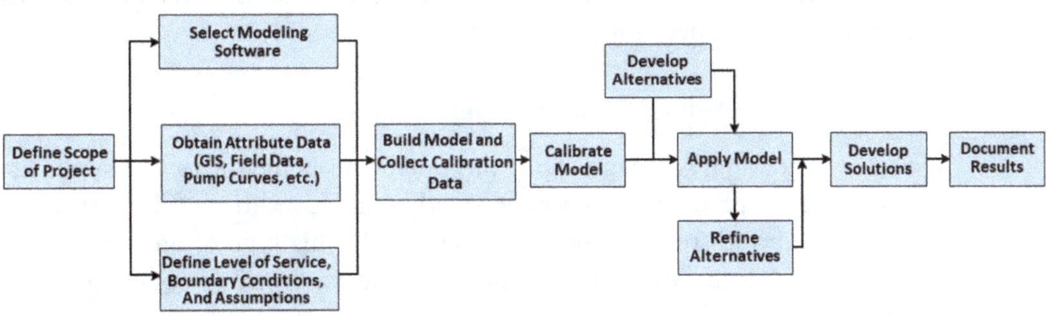

FIGURE 3.3 Hydraulic Model Development Workflow

response to storm events as well as normal dry weather flows. The typical RDII sources within a collection system contribute flows from multiple pathways. This makes quantifying RDII challenging and requires flow and rainfall monitoring data to improve RDII quantification and to better understand system performance during various rainfall events. Analyzing flow and rainfall data will help to determine where the greatest effects are within the system during a storm event, help to target high rate RDII locations, provide data necessary to evaluate system improvements to eliminate capacity-limited infrastructure, and facilitate identification of target areas for system rehabilitation.

3.1.3 Climate Change Effects

Effects of climate change to an existing collection system will vary upon the geographical location. Climate change effects can include increased yearly rainfall totals, increased storm intensities, and increased groundwater levels. Collection systems located in coastal areas can be affected by high groundwater tables, at or above sewer inverts, that result in baseline infiltration. Baseline infiltration effects occur during both dry weather and wet weather conditions and can result in increased flows to the system on a 24-hour basis, often greatly exceeding total volume of weather-related peak flow events. The resulting effect is reduced system capacity and increased treatment costs. Piezometers can be installed in areas with high groundwater tables to better monitor and track the groundwater conditions through seasonal changes.

3.2 Predictive Hydraulic Models

Predictive hydraulic models include the addition of inflow of infiltration rates resulting from a defined rainfall event or events for a specific level of service or regulatory required design storm and can include forecasted storm events for a given area. Information derived from predictive hydraulic models can be used to better develop response plans, identify system upgrades, and implement operational modifications.

3.2.1 Steady-State Hydraulic Models

Steady-state analysis determines the operating behavior of the system at a specific point in time or under steady-state conditions (flowrates and hydraulic grades remain constant over time). This type of analysis can be useful for design of new or expansion projects, typically concerned with peak flow conditions.

3.2.2 Dynamic Hydraulic Modeling

Dynamic hydraulic modeling generally implies the simultaneous solution of the St. Venant equations conserving both mass and momentum at several points along each pipe. When backwater conditions are present, or multiple flow paths exists such as in regulators or having more than one outlet in a manhole, a dynamic model is appropriate. In such instances, steady-state models would be inadequate to accurately simulate the system. Additional advanced analyses include computational fluid dynamics and two-dimensional simulations that integrate the mutual effects on drainage networks, rivers, and floodplains.

3.2.3 Real-Time Hydraulic Modeling

Real-time hydraulic modeling software is now capable of integrating hydraulic models, Supervisory Control and Data Acquisition (SCADA) historians, and GIS programs to provide near-real-time updates on system performance. These models incorporate radar and rain gauge data to forecast the effect of rainfall on collection system performance, flooding, and probability of overflow events.

4.0 STOPPAGE AND OVERFLOW RESPONSE

Despite best efforts, collection and conveyance systems will remain susceptible to SSOs that can result from a multitude of reasons that are unforeseen or less predictable. To address these occurrences, development of an SSO response plan will help mitigate the magnitude and duration of the SSO. Response preparedness should include availability of staff, equipment, and materials.

4.1 Emergency Response Times

Improving response times for SSOs can be achieved using the information, tools, and technologies described above. For example, the collected data stored in a GIS database that was used to produce a heat map graphically represents where frequent or chronic SSOs occur, thus allowing field crews to target response and control resources to those areas and causes. Further, if the data on the weather conditions at time of occurrence is collected, this information can be used as a predictive tool to verify availability of staff, materials, and equipment as a means to help improve response times when similar weather is predicted. An accurate GIS database combined with a hydraulic model will better serve to forecast the performance of a collection system for a variety of environmental conditions. These technologies help to facilitate the sharing of system operational knowledge with all management and operations staff.

4.2 Emergency Response Preparedness

Sanitary sewer overflow events rarely occur at convenient times. Being adequately prepared serves to reduce the system downtime and duration of the SSO. Preparedness can include material availability, access to the necessary equipment, and staff availability. Utilization of a computerized maintenance management system (CMMS) will help in the proactive maintenance critical item availability needed for response and corrective actions.

4.2.1 Repair Material Inventory

Defining the repair materials necessary to maintain in stock can be achieved using system record drawings, an accurate GIS database, or a system hydraulic model. The level of preparedness may vary between utilities, but at a minimum, the inventory should include frequently used items. These items often include pipe, repair clamps, repair sleeves, and repair tools. A fully populated GIS database will have the needed existing asset data required to define critical items. The inventory can be based on a simple assessment of the existing system and based on a percentage of quantity of the various system assets with an assumption of repairs per month. However, utilization of an accurate GIS database and CMMS that includes system failure records and material usage, the repair inventory can be more focused on historical system performance and material consumption.

4.2.2 Repair Equipment Inventory

Like the material inventory, the equipment inventory should be relative to the types and sizes of system repairs that are most frequent. Considerations for equipment needs should include the pipe diameters, buried infrastructure depths, dewatering needs, and emergency bypass pumping requirements. Personnel responding to the SSO event should be equipped with proper personal protective equipment for handling potentially hazardous materials during adverse weather events. Additionally, access to temporary traffic control equipment is necessary to protect the site and personnel during the response.

4.2.3 Staff Availability and Training

Most utilities optimize staffing based on normal operating procedures; however, consideration should also be made based on forecasted storm events. Using the historical data collected and stored in GIS or a CMMS, more accurate predictions can be made to help ensure adequate staff is available to address system issues.

Staff assigned to SSO emergency response should be trained in proper response procedures. Field personnel should be prepared to estimate the volume of wastewater released during an SSO event and to complete field reporting forms with the information required under the state's SSO notification and reporting requirements. States may have follow-up sampling and monitoring requirements for pollutants that requires proper sampling protocols and chain-of-custody forms that must be completed by field personnel.

5.0 REFERENCES

Water Research Foundation. (2003). *Reducing peak rainfall-derived infiltration/inflow rates—Case studies and protocols* (Project No. 1331). https://www.waterrf.org/research/projects/reducing-peak-rainfall-derived-infiltrationinflow-rates-case-studies-and

6.0 SUGGESTED READINGS

American Water Works Association. (2017). *Computer modeling of water distribution systems* (4th ed., Manual of Water Supply Practices M32).

Collection Systems Committee. (2017). *Fact sheet—Sanitary sewer systems: Rainfall derived infiltration and inflow (RDII) modeling* (WSEC-2017-FS-001). Water Environment Federation, https://www.wef.org/globalassets/assets-wef/direct-download-library/public/03---resources/wsec-2017-fs-001-rdii-modeling-fact-sheet---final.pdf

National Association of Sewer Service Companies. (2018). *Trenchless Technology and Asset Management.*

U.S. Environmental Protection Agency. (2005). *Guide for evaluating capacity, management, operation, and maintenance (CMOM) programs at sanitary sewer collection systems* (EPA 305-B-05-002). Office of Enforcement and Compliance Assurance. https://www3.epa.gov/npdes/pubs/cmom_guide_for_collection_systems.pdf

U.S. Environmental Protection Agency. (2007). *Computer tools for sanitary sewer system capacity analysis and planning* (EPA/600/R-07/111). Office of Research and Development. https://cfpub.epa.gov/si/si_public_record_Report.cfm?Lab=NRMRL&dirEntryID=184303

U.S. Environmental Protection Agency. (2008). *Review of sewer design criteria and RDII predictions* (EPA/600/R-08/010). Office of Research and Development. https://cfpub.epa.gov/si/si_public_record_report.cfm?Lab=NRMRL&dirEntryId=188284

U.S. Environmental Protection Agency. (2012). *SSOAP toolbox enhancements and case study* (EPA/600/R-12/690). Office of Research and Development. https://nepis.epa.gov/Adobe/PDF/P100G00U.pdf

4

Asset Condition Assessment

Phil Hubbard, PE & Christene Mitchell, PE

1.0 PIPE CONDITION ASSESSMENT

There are multiple means to assess the condition of your pipelines, but the ability to use these technologies and the quality of the results depend on many factors including pipe use, pipe type, and pipe flow. This chapter discusses the multiple means, their value, and their limitations.

When evaluating capabilities of condition assessment technologies, pipe types include but are not limited to, nonferrous, specifically polyvinyl chloride (PVC), high density polyethylene (HDPE), fiberglass, reinforced concrete pipe (RCP), prestressed concrete cylinder pipe (PCCP), asbestos, and vitrified clay; and ferrous, specifically cast iron, ductile iron, and steel. When the term "any" is used in context to pipe type, it means the technology can be utilized for condition assessment of any of those pipe types listed.

1.1 Types of Pipe Assessment Technologies

This subsection provides general knowledge of available condition assessment technologies for pipelines. For each condition assessment technology, you will learn about how it works and general limitations. At the end of this subsection are tables (Table 4.1 and Table 4.2) listing pipe types applicable for each condition assessment technology.

1.1.1 Camera Systems

There are multiple camera systems designed for sewer collection systems. They allow the user to review the condition of the interior of their systems. In all cases, the cameras can only show conditions outside of water, so camera systems are limited to gravity systems with low flow, or out-of-service siphons or force mains. Typically, flow can be managed in gravity systems to utilize camera systems. Considering this and that the cameras provide the user the ability to review inside their pipes, cameras are the most popular technology for gravity system condition assessments. The commonly used camera systems in sewer collection systems are reviewed in the following subsections.

1.1.1.1 Closed-Circuit Television

Closed-circuit television (CCTV) inspection is a video inspection. A CCTV camera is mounted onto a tethered remote-controlled crawler or on a float/skid that is pulled through the line (device) and transmits high-resolution images to an above-ground computer, which is typically mounted in a vehicle designed for CCTV inspection and other related activities. Camera systems can be fitted to an array of vehicles to help mobilize and commensurate to the conditions of the site as necessary. Continuous video is recorded as the device runs through the pipe. The device receives separate actions than the camera. The device can be sped up, slowed down, stopped, and backed up. The camera can be fixed or be one with pan and tilt capabilities, allowing the user more capability to explore specific areas in the pipe (i.e., defects, laterals, joints). Some devices also have the ability to elevate the camera to allow recentering of the lens if diameter changes occur midsegment. The optimal camera speed is set by the National Association of Sewer Service Companies (NASSCO) Pipeline Assessment Certification Program (PACP) guidelines and is currently set to be 30 feet per minute to minimize camera blur. The camera can be rotated and zoomed to see parts of the pipes of greater interest.

Limitations of this technology include: (a) the pipe must be clean enough to allow the device to drive through it; (b) depth of flow in the pipe needs to be low enough to prevent submergence of the camera system selected; (c) condition assessment can be limited to what is seen above the waterline; and (d) the device tether limits the length of pipe the device can run through.

Pipe cleaning is normally sufficient to clear pipes enough to allow the device to pass. The minimum diameter needed for CCTV is 150 mm (6 in.). The flow limitation is overcome by inspecting during low flow periods; plugging flow; providing assistance with a combination vacuum and pressure (CVP) hose to flush or vacuum the line; or redirecting flow via bypass. However, as stated, the flow limitation restricts use of this technology to gravity systems and out-of-service siphons and force mains. The limitation of condition assessment only above the waterline is not normally a concern depending on the diameter of the pipeline. A user can get a good idea of the pipes' overall condition by just seeing above the pipe springline, the top 50% of the pipe; however, for pipes 300 mm (12 in.) and smaller, it is suggested full pipe diameter (100%) be visible via the use of CVP. When the full pipe is not visible, the user should consider internal and external environments that may cause an isolated defect under the water (i.e., increased/decreased pH in the wastewater that could corrode the pipe, a ferrous pipe surrounded by corrosive soils, or internal tuberculation of cast iron and ductile iron pipe). Regarding the tether limitation, most utilities have accessible manholes that are close enough to allow a standard CCTV system to work. Most CCTV systems have tethers as long as 457 m (1500 ft) as standard but can utilize tethers up to 5 km (3 mi) in length using Kevlar (e.g., a heat-resistant and strong synthetic fiber) reinforced cabling for larger diameter pipe inspection.

The interior of force mains or fully submerged is much darker than partially submerged vitrified clay or concrete gravity mains, so it is harder to see interior conditions, especially when the pipe itself is a darker material (i.e., Orangeburg, cast iron, ductile iron), even with additional lighting on the camera. Wastewater in full pipe environments will result in the generation of hydrogen sulfide (H_2S) gas where bacteria can create a darkened slime layer on the pipe that will impede clear images of the pipe inspection.

Camera resolution has increased from low video graphics array (VGA) 640 × 480 pixels to higher definition phase alternating line (PAL), 720 pixels, to even higher 1080 pixels and 4000 megapixels having varying screen ratios of 4:3 for earlier lower resolution iterations to widescreen 16:9 orientation for the latest higher resolutions.

Closed-circuit television platform setups typically transition from small to large diameter setups at around the 920 mm (36 in.) diameter where lens resolution at 1080 pixels or greater resolution and brighter forward-facing lighting that can dim during panned environments are swapped out accordingly. For larger diameter inspections, the focal length will be much greater, requiring the use of more effective lighting systems and high definition, high resolution camera systems to enable the viewer to observe and confidently assess defects and resultant condition in the perspective view. The CCTV camera must have its own illumination source, where the lighting shall be intrinsically safe, fixed in intensity to allow the whole pipe surface to be illuminated without shadow, and should be capable of extending light twice the diameter beyond the camera lens focal length. Distances of defects must be observed at the side periphery or right and left edges of the screen where the image plane and focal point meet and logged accordingly. Observations measured from the obvert or invert will introduce errors, especially if the camera is not centered in the circular or egg-shaped pipe.

A CCTV condition assessment includes defect condition scoring that is typically exported as a Microsoft Access database, a paper exported report, and recorded CCTV video. Condition scoring can be based on a utility's preference; however, the industry standard is NASSCO's method of scoring criteria for the PACP, laterals (LACP), and manholes (MACP) in North America (NASSCO, 2021).

1.1.1.2 Panoramic Camera System

Panoramic camera systems are similar to CCTV camera systems but use side wall scanning technologies. Like CCTV, the panoramic pipe inspection system is a camera system attached to a tethered remote-controlled device and therefore have the same limitations as per the CCTV system. Side wall scanning technologies typically use two cameras contained within the unit, taking multiple 180° still images (front and back) where software will stitch the thousands of images together to create a 360° pan and tilt video environment. This method of side wall scanning technology enables the viewer to autonomously pan and tilt within the viewing software at varying distances at their own discretion. Subsequently, having this 360°

focused view eliminates the concern that the camera operator may have overlooked an area that the end user may need for condition assessment. In larger pipes, however, lower resolution (first generation) variants of this technology have inhibited the viewers' ability to correctly discern levels of pipe wall loss because of H_2S attack. The technology produces photographic images and curves them to the predetermined shape of the asset. The "flat" two-dimensional image is placed into a three-dimensional environment that may inhibit the assessor's ability to discern the corroded pipe wall loss depth.

There are also versions designed specifically for manhole condition assessment, which is essentially the camera device on a cable system that can be easily lowered into the chamber without the use of physical entry to the chamber.

1.1.1.3 Zoom Camera

Zoom camera is a term that refers to a high-resolution camera attached to a pole, which allows a user to quickly and easily lower the camera into a manhole and capture photographs and video of the manhole and/or through pipelines. It uses a technology that allows the zoom functionality to link with the probable distance it is viewing within the pipe. This method can be used to provide a complete condition assessment of a manhole, but pipe inspection is often limited to the zoom capability and the pipe environment because many defects and features observed within the pipe may obscure further zoomed observations. Consequently, reverse inspections are often required to complete a full inspection. Many utilities use this technology for reactive spot inspections for operational issues or to help further develop a defined inspection program using in-line CCTV camera tractor units.

1.1.2 LASER and LiDAR Profiling

A light amplification by stimulated emission of radiation (LASER) and light detection and ranging (LiDAR) profiling inspection provide both a visual representation of the pipe and fine measurements inside the pipe. This allows the user to view a more accurate determination of pipe geometry (features and defects) above the fluid level. Both technologies use lasers to create the profile and take measurements; and both devices mount atop of a tethered remote-controlled device that are run through the pipe to collect data.

The difference between LASER and LiDAR is how the laser light is captured to provide the profile and measurements. In LASER profiling, the laser light is projected 360° around the pipe to form a ring around the interior of the pipe and a camera on the device captures the ring. Measurements are calculated based on what the camera sees. In LiDAR profiling, the laser is projected 360° around the interior of the pipe, and the profile is developed and measurements taken based on the time it takes for the laser to hit the pipe wall and return back to the device, much like sonar. Deliverables can produce internal diameter cross sections and deflection graphs to quantify internal pipe wall material loss/gain or deformation (ovality and deflection) at given frequencies along the length of the sewer segment where surface corrosion measurements can be accurate to 5 mm at 3 m in 1220 mm (48 in.) pipes and larger. Light amplification by stimulated emission of radiation scans can produce a point cloud three-dimensional environment having distance points of 10 mm in the transverse direction and 40 mm in the longitudinal direction, which is more than adequate to determine pipe shape and condition when reviewing in conjunction with the CCTV video.

Advancements have been made in LiDAR profiling to now offer 3D LiDAR. Three-dimensional LiDAR creates a three-dimensional representation of the inside of a structure. To get this third dimension, the unit must remain steady, so it cannot travel through the pipe. This means it can only create three-dimensional profiles in the laser light range of where it is sitting.

The limitations of LASER and LiDAR profiling is it cannot capture data below the waterline in sewer systems.

1.1.3 Transient Pressure Wave Systems

Transient pressure wave systems are a means to assess the condition of a pipe by creating a controlled transient pressure wave in a pressurized water/wastewater system and measuring the receiving pressure wave and reflections of the pressure wave. Currently, this technology is only offered in the United States by Hydromax USA and is called Pipeline Condition Assessment Technology (p-CAT). Hydromax USA entered into an exclusive licensing agreement with Detection Services in Australia, which has been successfully using this technology for over a decade in Australia and New Zealand. This technology is able to measure the pipe wall thickness, change in internal wall thickness (i.e., spalling or change in pipe type), obstructions, and air/gas pockets. Location accuracy is +/− 10 m (+/− 33 ft).

As implied in the previous paragraph, this technology can only be used on force mains. Its main advantages are it can be done without taking the force main out-of-service (in fact, it requires it remain in service), and it requires no destructive changes to the force main to use. It can take advantage of existing air release valves to be pressure transmission and receiving locations as long as they are approximately 1.6 km (approximately a mile) apart. If additional pressure transmission/receiving locations are required, the equipment can be installed on the exterior of the pipe without having to cut/drill into the pipe. The minimal system pressure requirement is a limitation that tends to exclude the use of this technology on many force main systems. Transient pressure wave technology currently requires system pressures above 210 kPa (30 psi).

1.1.4 Remote Field Technology

Remote field technology (RFT) refers to devices that collect data in a submerged water system without taking it out-of-service. There are both free swimming and tethered devices. These devices take advantage of one or more of the other technologies that will be discussed, such as magnetic flux leakage and ultrasonic testing; however, their ability to be inserted into a submerged pipe and collect data "hands-free" is what categorizes these devices as RFTs. There are multiple RFTs on the market that are designed to assess condition in different ferrous and concrete pipe types. RFTs can capture obstructions, pipe wall anomalies, pipe wall thickness, and air/gas pockets. Current examples include Xylem's PipeDiver and PICA's multiple SeeSnake models (SeeSnake, SeeSnake-Chimera, SeeSnake RAFT, SeeSnake-EMIT).

There are challenges associated with using RFTs. First, they have system velocity and pressure constraints, which often requires changing flow conditions in the pipe to be assessed, and in some cases, precludes the use of specific RFTs in specific pipes. Another challenge associated with RFTs is getting the device in and out of the pipe without getting it stuck. Often a pump station wet well or water resource recovery facility headworks can be the extraction location for force mains, but if an insertion opening for condition assessment technologies doesn't already exist on the system, it will need to be added. Also, ensure the RFT is able to easily travel through the pipeline system before insertion by studying the pipeline plans for potential snags (i.e., pipe size changes, appurtenances, bends). Getting one stuck can be costly to retrieve.

1.1.5 Sonar Technologies

This section covers all technologies that utilize sound waves to obtain data on condition assessment. Sound waves are emitted from these sonar devices, and pipe conditions are determined by the sound characteristics reflected to the device or captured by other devices.

1.1.5.1 Sonar Profiling

Sonar profiling utilizes a device that is mounted to a tethered remote-controlled device and run through the pipe below the water line. Sonar profiling works by emitting sound waves and analyzing the time the sound takes to hit the pipe wall and reflect back. Longer time than expected tells the device there is degradation on the pipe wall. Shorter time than expected tells the device there is buildup of debris. Sonar profiling can also provide a picture of the current shape of the pipe, which is valuable if the person assessing the pipe is trying to understand if load on the pipe is causing ovality.

Sonar scans provide, in general, two forms of information that include the amount of debris in the pipe and the interior shape of the pipe below the water level. Debris (outside of encrustations on the pipe wall) located on the pipe invert results in a truncated shape of the normally circular pipe cross-section data with a cord running across the profile circle that graphically represents the debris surface. Scans are conducted continually with a sonar head rotating on an axis parallel to the longitudinal direction of the pipe. Pipe restriction is calculated by using the assumed area of the pipe cross section based on the record drawing pipe diameter and the cross-sectional area of the debris. The area occupied by the debris is calculated from an area bounded by the debris surface location, as determined from the sonar scan, and the inferred pipe invert, assuming a circular pipe shape. Sonar scans are more commonly used to determine the amount of debris and level of restriction in a pipe and less commonly used to assess pipe geometry and inherent structural ramifications of the same.

Sonar profiling is typically performed in gravity sewer systems when flow cannot be managed and in siphons that remain submerged. Depending on pipe diameter, the device can either be mounted on tracker/crawler unit or floated platforms or mobilized independently to facilitate inspections within pipe diameters as small as 200 to 300 mm (8 to 12 in.); however, less spurious data can be achieved within larger diameter pipelines.

1.1.5.2 Ultrasonic Testing

Ultrasonic testing (UST) refers to the use of high frequency sound waves to detect the thickness of ferrous pipes. This technology is utilized on some RFTs. For the purpose of this subsection, UST is referring to the use of a hand-held device with a probe that the user must place directly on the spot on the pipe to measure thickness.

Ultrasonic testing can provide high accuracy of spot thickness of a ferrous pipe when calibrated to the exact material being tested. The device is dependent on the input of the sound speed of the material being tested, so if the sound speed input is not correct, the thickness output will not be correct. For instance, users need to be aware all ductile iron is not the same, because of differing classes of pipe, so a ductile iron sample from one class of pipe may not work to calibrate the device for another class of ductile iron pipe. This is important to understand, yet unfortunately, overlooked by many.

Not having samples or sound speeds to calibrate the UST is one limitation. Another is the fact it can only provide spot thicknesses at the location tested. The section of the pipe to test must be exposed. It is difficult to get an overall picture of the condition of the pipe using UST, so it is best to use this technology in locations you need to confirm/deny pipe degradation. A third limitation is the spot to be tested must be smooth metal. Any coatings need to be removed, and pipe dimples sanded down. Users should be aware of the depth a pipe is sanded, as it can affect the results.

1.1.5.3 Guided Wave Technology

Guided wave technology (GWT) produces sound waves on the outside of a ferrous pipe and measures it along the pipe to capture interior pipe degradation to include percent pipe wall loss. It is also able to capture leaks in the pipe. This secondary benefit is currently the sole benefit for wastewater systems. Unfortunately, this system is not yet able to gather accurate enough data in a wastewater system to produce reliable pipe wall loss.

For those interested in using the technology for leak detection on ferrous pipes, you will need access to pipe appurtenances for sound transmission/receiving (i.e., air release valves) between 90 and 180 m (300 and 600 ft) apart.

1.1.5.4 Sonic/Ultrasonic Inspection

Sonic/ultrasonic inspection is a means to assess the condition of PCCP pipelines by directly affecting the pipe wall producing sound waves, and then measuring the time and frequency those sound waves have traveled a distance in the pipe wall. This technology can determine if there is any degradation or loss of strength or integrity of the PCCP. Specifically, sonic/ultrasonic inspection can identify if broken prestressing wires have slipped and core concrete is delaminated from the cylinder.

This technology requires the pipe area to be exposed and available to directly apply to the outside of the pipe. Like UST, the pipe condition is only determined at the spots the device is applied; therefore, an overall condition of the pipe is not practicable with this technology. It is best to use this technology in locations where confirmation/rejection of possible pipe degradation conditions is needed.

1.1.6 Electromagnetic

This section covers all technologies utilizing electromagnetic waves to obtain data on condition assessment. Electromagnetic waves are emitted from these devices, and pipe conditions are determined by the signals reflected back to the device or captured by other devices.

1.1.6.1 Pipe Penetrating Radar

Pipe penetrating radar (PPR) uses the same technology as ground penetrating radar (GPR) on pipes. Like GPR, PPR is emitting electromagnetic waves and measuring the reflected signals. Pipe penetrating radar devices are specifically designed to be inserted in sewer systems to provide condition assessment. These devices can provide wall thickness (or percent wall loss) of nonferrous pipes, the presence/absence of rebar, and detect the presence of voids outside nonferrous pipe. (Fiberglass pipe, although a nonferrous material, is not verified for use of this technology.)

The use of electromagnetic technology to perform condition assessment is utilized in some RFTs, and the term, PPR, does not refer to these applications. The term, PPR, is specifically used when referring to this stand-alone technology. Pipe penetrating radar technology can be affixed to a remote-controlled device and run through a pipe to collect data internally. Pipe penetrating radar technology is often limited by the device it is attached. Pipe penetrating radar can collect data submerged, if the device can get through the pipe, it can collect data.

1.1.6.2 Broadband Electromagnetics

Broadband electromagnetics (BEM) utilizes magnetic technology to assess the condition of ferrous pipes. A pulse eddy current is applied directly to the pipe, and the receiving current allows the user to determine the density of the ferrous pipe and identify flaws on either side of the pipe wall and percent wall loss. Broadband electromagnetics is used in some RFTs; however, in this subsection BEM refers to the use of this technology on the outside of a pipe to assess conditions. Currently, external BEM is proprietary to Elite Pipeline Services.

Because this technology requires the current to be directly applied, the pipe must be exposed for use. Therefore, like UST and sonic/ultrasonic inspection, this technology is best used in locations needed to confirm/deny pipe degradation. This technology does have an advantage over UST, it does not need to be in direct contact with smooth ferrous material. Coatings and corrosion can remain on the pipeline during testing.

1.1.6.3 Electromagnetic "Bracelet"

Electromagnetic "bracelet" is a device made up of multiple high-resolution sensors on a ring that is wrapped around the outside of a ferrous pipe to detect internal degradation (percent wall loss, pits, cracks, and graphitic corrosion). The device emits a low frequency electromatic field that goes through the pipe wall. Changes in the pipe wall thickness will result in changes in the electromagnetic field, which the bracelet sensors detect.

Like BEM, the pipe must be exposed for use, and coatings do not need to be removed. Because it is a spot condition assessment, this technology is best used in locations needing to be confirmed/deny degradation. The electromagnetic "bracelet" can be adjusted to fit any pipe size above 150 mm (6 in.).

1.1.6.4 Magnetic Flux Leakage

Magnetic flux leakage is a process of highly magnetizing a ferrous pipe and measuring magnetic flux leakage that is caused when there are flaws in the material, such as corrosion. The amount of leakage can be analyzed to determine the precise wall loss of the pipe at that location.

There are both internal and external magnetic flux leakage devices on the market. Use of the external allows the pipe to remain in service; however, the pipe being tested must be exposed. Like many of the other technologies, the external devices only provide condition assessment at those locations exposed, so it is best used to confirm/deny pipe degradation (percent wall loss). There are various internal magnetic flux devices. There are some that require the pipe to be out-of-service, and some are free swimmers. The chosen device will dictate whether the pipe sizes can be assessed from 150 mm to 2 m (6 to 80 in.).

1.1.7 Acoustic Fiber Optic Monitoring Systems

An acoustic fiber optic monitoring system is a system installed inside, along a pipe system, typically during construction, that will provide continuous monitoring of the sound along the pipes. Currently, it is only used for PCCP to listen for the breaking of the prestressed steel wires. One needs to weight the use of this device in sanitary force main systems, with the possibility of debris collecting on the fiber optic line.

In water systems, the loss of prestressed wires greatly reduces the strength of the pipe because of the high pressures typically found in water systems. For sewer force mains that normally do not experience high internal pressures, the knowledge of prestressed wires breaking lets the utility know that the outer layer of the pipe has deteriorated, and the useful life of the pipe is reduced.

1.1.8 Acoustic Leak Detection Systems

Not to be confused with sonar, there are devices that can be put in pipes that use acoustics to listen to the sounds in the pipe and identify the sounds of leaks. There are both free-swimming and tethered devices. Depending on the device, the acoustic systems have different means to provide their location along a pipe as the systems collect acoustic data. Some of these devices can also identify air pockets, which could be sources of internal corrosion or capacity restrictions. As these devices are sensitive to the sound of air, acoustic systems are limited to submerged conditions, such as force mains.

Acoustic leak detection systems include Pure Technologies's SmartBall and Sahara, and PICA's Recon+.

There is another form of acoustic leak detection that is similar to GWT, but these technologies only claim to find leaks and do not offer the ability to assess wall thickness. Similar to GWT, devices are attached to the outside of the pipe or appurtenances, such as valves, and sound is transmitted and received. The change in sound speed lets the user know there are leaks. Unlike GWT, when using this technology to identify only leaks, pipe materials are not limited to ferrous only. This is because the sound they are listening for is traveling along the water column and not the pipe wall. Echologics's LeakFinder-ST and SebaKMT's Correlux C-3 are examples of this type of acoustic leak detection system.

It should be noted that the identification of leaks is an assessment of the condition, but the use of these devices to identify air pockets is not. Air pocket data provides greater information on how/where to focus condition assessment.

1.2 Condition Assessment Technologies Comparison

Because of the ample access gravity systems have, technologies that provide clear images and profiles of the condition of the interior of the pipes are often available and utilized. These technologies are camera systems, LASER/LiDAR profiling, sonar profiling, and PPR. Table 4.1 provides a summary of pros versus cons of these technologies for use in gravity mains. To learn more about these technologies, refer to Section 1.1, Types of Pipe Assessment Technologies.

When deciding which condition assessment technology to use on a force main, an understanding is needed of the technologies' limitations regarding pipe type, pipe size, whether it is an external or internal technology, and if the force main can remain in service. Table 4.2 provides the quick summary of these considerations. To learn more about these technologies, refer to Section 1.1, Types of Pipe Assessment Technologies.

2.0 PIPE APPURTENANCES CONDITION ASSESSMENT

2.1 Gravity System Laterals, Manholes, and Siphons

Technologies have been developed to easily assess gravity system appurtenances, specifically laterals, manholes, and siphons. This section will review these technologies.

TABLE 4.1 Comparison of Typical Gravity System Technologies (Reprinted with permission from HRSD)

Technology	Pros	Cons
Camera Systems (specifically CCTV and Panoramic)	• System is readily available (CCTV) • You can actually see the condition of your pipe • Standard condition coding system available • Can be used on any pipe type	• Unable to see pipe condition underwater • Degradation is left to interpretation of the viewer
LASER/LiDAR Profiling (2D)	• Ovality and degradation is measured • Can be used on any pipe type	• Unable to measure pipe underwater • Profiles are not accurate in pipes over 900 mm (36 in.)
Sonar Profiling	• Can provide a representation of conditions underwater • Can be used on any pipe type	• Limited to learning pipe wall loss and debris buildup
Pipe Penetrating Radar (PPR)	• Provides pipe wall thickness • Can provide the presence/absence of rebar • Provides location of voids outside of the pipe	• Cannot be used on ferrous pipe and PCCP, and use on fiberglass is not verified

2.1.1 Laterals Inspections

Gravity sewer laterals can be inspected internally using camera systems. There are two camera systems typically used to understand the condition of laterals. One is the push camera. This is a common camera system used by plumbers and utilities to identify issues in the laterals. This is simply a camera at the end of a long stiff cable the user can insert into any cleanout and "push" through the lateral to inspect. The other is the CCTV lateral launch. This system also has a camera at the end of a long stiff cable. The difference is the camera and cable are attached to a remote-controlled device and placed in the gravity main to explore the lateral starting from the lateral connection to the gravity main. The device is typically placed in the pipe and

TABLE 4.2 Comparison of Typical Force Main Technologies (Reprinted with permission from HRSD)

Technology	Pipe Type	Pipe Diameter	Internal/External Inspections	Remains in Service
Camera Systems	Any, but visibility is difficult in ferrous pipes	> 150 mm (6 in.)	Internal—but not submerged	No
LASER/LiDAR Profiling	Any	> 150 to 200 mm (6 to 8 in.) < 900 mm (36 in.) without loss of accuracy	Internal—but not submerged	No
Sonar Profiling	Any	> 200 to 250 mm (8 to 10 in.) cleaned	Internal	Potentially
Transient Pressure Wave Systems	Ferrous Asbestos	200 to 900 mm (8 to 36 in.)	External	Yes
Remote Field Testing (RFT)	Ferrous RCP PCCP	Varies	Internal	Yes, if access ports available
Ultrasonic Testing (UST)	Ferrous	Any	External—pipe exposed	Yes
Guided Wave Technology (GWT)	Ferrous	Any	External	Yes
Sonic/Ultrasonic Inspection	PCCP	Any	External—pipe exposed	Yes
Broadband Electromagnetic (BEM)	Ferrous	Any	External—pipe exposed	Yes
Pipe Penetrating Radar (PPR)	PVC HDPE RCP Asbestos Vitrified clay	200 to 1200 mm (8 to 48 in.)	Internal	Yes
Electromagnetics "Bracelet"	Ferrous	> 200 mm (6 in.)	External—pipe exposed	Yes
Magnetic Flux Leakage	Ferrous	Any	External—pipe exposed	Yes
			Internal	Depends on device selected
Acoustic Fiber Optic Monitoring	PCCP	Any	External—installed at construction	Yes
Acoustic Leak Detection Systems	Any	Varies	Internal	Yes, if access ports available

run in a reverse setup, meaning the device is moved upstream through the pipe, which allows the camera to more easily enter laterals that enter the pipe on a diagonal. Once at the lateral, a cable with a camera is extended into the lateral from the gravity main for inspection. The launched camera typically can be extended to the face of the house. Like camera system inspections of sewer mains, NASSCO has a standard condition scoring for laterals (LACP) that can be used to score lateral condition.

It should be noted that laterals extend into private property and at that point are private systems, and the utility may need permission from the property owner to assess the private portion of the lateral. A cleanout is typically installed at the property line to distinguish the public and private systems, also called upper lateral (generally private, on private property) and lower lateral (generally public, within the municipal right-of-way). There are also cases when the utility only has authority to maintain the sewer main, and property owners are responsible for the entire lateral, even within the municipal right-of-way. In those cases, if a utility wanted to assess the lateral, the utility may request permission to assess the lower lateral and upper lateral (100% of the lateral).

2.1.2 Manhole Inspections

Condition of manholes can be performed without entry using camera systems. There are specific panoramic cameras available that capture multiple still frames as the user lowers it into the manhole to create the 360° focused view. Zoom cameras are also popular for condition assessment of manholes. Both camera systems are suitable for using NASSCO's condition assessment scoring for manholes (MACP).

Another option for condition assessment of manholes is 3D LiDAR. 3D LiDAR can be set up and stabilized in a manhole to create a three-dimensional profile of the manhole.

2.1.3 Siphon Inspections

Siphons are essentially depressed gravity sewers. Camera systems and LASER/LiDAR profiling will not work in these locations unless out-of-service and drained, but sonar profiling has proven successful in these areas. External condition assessment technologies stated above can also be used.

2.2 Other Appurtenances and Technology Considerations and Limitations

There is so much more to our collection systems (valves, air vents, clamps, flanges), and while they may seem ancillary, experience shows that failure of these appurtenances can cause significant damage. Examples include:

- Valves that do not close during a failure, causing greater release of wastewater;
- Valves that are stuck closed, causing a wastewater backup;
- Air vents not operating, causing corrosive gas to build up and degrade the pipe, or reduced flow capacity and increased pipe pressure; and
- Bolts on clamps and flanges corroding away breaking open the clamp/flange.

Although there are currently no technologies to assess the condition of these appurtenances, it is important not to ignore appurtenances. Taking advantage of other tools and technologies can be used to, at least, assist in prioritizing appurtenance condition assessments. Such other tools and technologies include:

- Resistivity studies of the environments along sewer mains to identify areas with higher risk of pipe and appurtenance corrosion; and
- Risk modeling to identify high-risk sections of your system and critical valves.

3.0 REFERENCES

National Association of Sewer Service Companies. (2021). Resources. https://www.nassco.org/resources/

4.0 SUGGESTED READINGS

Aanderud, J. (2009, April 1). Steps to consider when establishing CCTV inspection programs. *WaterWorld*. https://www.waterworld.com/home/article/16193571/steps-to-consider-when-establishing-cctv-inspection-programs

BEM Pipe Inspection. (2015). About BEM inspection. https://www.beminspection.com

Bueno, S. (2020, February). Artificial intelligence: Pipe inspection's next big thing. *Trenchless Technology*. https://trenchlesstechnology.com/artificial-intelligence-pipe-inspections-next-big-thing/

Detection Services. (2021). Magnetic flux leakage. https://www.detectionservices.com.au/technologies/smart-cattm/

Echologics. (2021). ePulse pipeline condition assessment. https://www.echologics.com/services/condition-assessment/epulse/

Eddyfi Technologies. (2021). Magnetic flux leakage. https://www.eddyfi.com/en/technology/magnetic-flux-leakage-mfl

Farr, A. (2017, April 24). Condition assessment using transient waves. *Water Finance & Management*. https://waterfm.com/condition-assessment-transient-waves/

Gipson, B. (2012, February 15). A look at laser profiling. *Trenchless Technology*. https://trenchlesstechnology.com/a-look-at-laser-pipe-profiling/

Griffin, J. (2012, February). Panoramic pipe inspection for manholes. *Underground Construction*. https://ucononline.com/magazine/2012/february-2012-vol-67-no-2/features/panoramic-pipe-inspection-for-manholes

Griffiths, J. (2020, December 1). Sensor technology: 2D vs. 3D. *RedZone Robotics*. https://redzone.com/nr/sensor-technology-2d-vs-3d/

Gunn, E. (2009, July). SEWER: New ways of seeing. *Municipal Sewer & Water*. https://www.mswmag.com/editorial/2009/07/sewer-new-ways-of-seeing

Heim, D. (2008, June). Ring of light. *Municipal Sewer & Water*. https://www.mswmag.com/editorial/2008/06/ring-of-light

NDT Corporation. (2020). Sonic/ultrasonic inspection. https://www.ndtcorporation.com/capabilities/sonic-ultrasonic-inspection

Russell NDE Systems, Inc. (2021). Bracelet probe on force main. https://www.russelltech.com/Products/Bracelet-Probe-Uninsulated-Pipe.

SewerVUE. (2015). Revolutionary pipeline condition assessment. https://sewervue.com/pipe-penetrating-radar.html

U.S. Environmental Protection Agency. (2010). *Report on condition assessment of wastewater collection systems* (EPA 600/R-10/101). Office of Research and Development. https://www.wef.org/globalassets/assets-wef/3---resources/topics/a-n/collection-systems/technical-resources/epato592010reportonconditionassesmentofwwcs-1.pdf

Water Environment Federation & National Association of Sewer Service Companies. (2019). *Trenchless technologies and asset management* (3rd ed., WEF Manual of Practice). Water Environment Federation.

White, C. (2020, September 1). Sonar lets organizations inspect pipes with flow. *RedZone Robotics*. https://redzone.com/nr/sonar-tackles-pipes-with-flow/

5

Collection System Rehabilitation and Replacement Technologies

Christy Willson, PE; Dianne Sumego, PE; & Tim Sumner, PE

1.0 GRAVITY SEWERS

Wastewater flows down the pipe as a result of the energy created from the difference in elevation or the slope of the pipe. Gravity sewers are the pipes that accept wastewater connections from residences, businesses, and industries. In most communities, a majority of pipes conveying wastewater are gravity sewers. Because of insufficient funds, age, and deterioration over time, gravity sewers are often in need of rehabilitation. Fortunately, there are several different technologies available to rehabilitate gravity sewers with varying site conditions and project constraints. Structural strength and improved hydraulics typically result from rehabilitation.

1.1 Rehabilitation

Sewer rehabilitation can utilize multiple technologies that address repairing and rehabilitating failing pipes. Rehabilitation means the integrity of the pipe is restored to some level that extends the service life of the asset depending on the technique used. The following sections describe the technologies of cured-in-place pipe (CIPP), slip lining, spiral wound materials, and spray-on linings.

1.1.1 Liners

Liners are installed in sewers of various shapes and sizes to restore the structural stability of the pipe. Generally, liners are installed from manhole to manhole in one continuous piece but can also be short liners to address point repairs. Various technologies are available to accomplish pipe rehabilitation.

Some applications require the flow to be bypassed, and lateral connections need to be reinstated once the liner is complete.

1.1.1.1 Cured-in-Place Pipe

Cured-in-place pipe is one type of liner that is installed in circular pipe to structurally rehabilitate the pipe barrel using a resin impregnated felt or glass fiber material that can be cured using different methods depending on the pipe diameter and process preferred by the contractor. Liner material can be shipped to the site impregnated with resin or injected with resin on-site for larger diameter installations. Flow must be bypassed during installation. After the pipe has been cleaned and inspected, liner material is inverted (i.e., turned inside out) into the pipe and prepared for curing. An example of CIPP rehabilitation is shown in Figure 5.1. Whether the liner material can be installed through the existing manhole opening or if the frame and/or cone section has to be removed depends on the sewer size being rehabilitated. Cured-in-place pipe is generally appropriate for pipe diameters of 10 to 244 cm (4 to 96 in.) (Water Environment Federation [WEF] et al., 2009).

FIGURE 5.1 CIPP Rehabilitation (Reprinted with permission from the Metropolitan St. Louis Sewer District)

1.1.1.1.1 Steam Cured

Steam cure of CIPP is typically used for small- and medium-sized pipe because of the short cure time. Water is heated by a boiler on-site to greater than the boiling point to create steam and pumped into the liner and sealed for the curing process. Cure time depends on the pipe diameter, depth, groundwater level, and length of installation but is faster than using water because heat transfer is quicker.

1.1.1.1.2 Hot Water Cured

Hot water cure is typically used for larger diameter pipe, longer runs, and pipe that may be close to failure because the pressure is held constant, and the water temperature is easily controlled. When the liner has been inverted into the host pipe, the pipe is filled with water, heated up to the required temperature, and held until the liner is cured. Water is then slowly cooled down to prevent liner shrinkage. Water cures often require the use of tanker trucks to transport the water. Unfortunately, the water used in the process is wasted.

1.1.1.1.3 Ultraviolet Technology

Instead of using steam or water to cure the liner, ultraviolet light can be used. Using ultraviolet light to cure the liner requires the use of a glass-reinforced liner that is pulled into the sewer and inflated rather than felt material that is inverted. A light curing train is needed to cure the liner instead of steam or water. The ultraviolet curing process is quicker than using water and stream cure, and no water is wasted in the process.

1.1.1.1.4 Ambient Air Cured

Ambient air cured liners are cured using the air inside the liner and the temperature in the sewer at the time. No outside means are used to cure the liner. Curing a liner at ambient temperatures is possible, but generally requires several days compared to hours with steam or water cures.

1.1.1.2 Slip Liners

Slip lining involves inserting a smaller diameter pipe into a larger diameter host pipe to rehabilitate the pipe. The methods used to install the slip lined pipe vary depending on the host pipe material and the slip lining pipe. Pipe manufactured with no bell or low bell is installed either with a winch and cable arrangement (pulled) or jacked into place (pushed). Any annular spaces remaining after the slip lining process are filled with low strength grout. A reduction in cross-sectional pipe area typically results, and the new pipe follows the slope of the host pipe. Flow may need to be bypassed based on pipe size, number of upstream connections, and site conditions. Lateral connections will have to be reinstated.

1.1.1.2.1 Polyvinyl Chloride

Polyvinyl chloride (PVC) pipe can be slip lined into existing sewers to restore the integrity of the host pipe. Fusible PVC is a familiar choice for slip lining smaller diameter pipe because it can be fused on the ground in a continuous process, in long or short lengths. Appropriate pipe sizes are 102 to 914 mm (4 to 36 in.) (WEF et al., 2009).

1.1.1.2.2 High-Density Polyethylene

High-density polyethylene (HDPE) pipe can also be used for slip lining and is similar in preparation and installation to PVC. Long continuous lengths of HDPE pipe can be fused and pulled into the host pipe rather quickly. Appropriate HDPE pipe sizes for slip lining are from 152 to 1600 mm (6 to 63 in.) (WEF et al., 2009).

1.1.1.2.3 Fiber Reinforced Pipe

Fiber reinforced pipe (FRP) is typically used for rehabilitating larger diameter sewer and is installed one section of pipe at a time. Each pipe segment is pushed into place through an excavated access pit. Low profile pipe or pipe with concealed joints are pushed into place and the annular space filled with grout. Flow may or may not need to be bypassed. Fiber reinforced pipe diameter pipe range is from 102 to 4293 mm (4 to 169 in.) (WEF et al., 2009).

1.1.1.3 Geopolymer Spray

Geopolymer mortar is a high strength material that provides structural integrity and corrosion resistance, see Figure 5.2. This material can be sprayed on, spin cast, or troweled on to various shapes and sizes of sewers. The difference between geopolymer mortar and shotcrete (see Section 1.1.1.7) is the geopolymer has greatly reduced calcium hydroxide concentration, which reduces the potential for corrosion of host materials (Najafi & Gokhale, 2005). Spray-on coatings can be applied to pipes 762 mm (30 in.) or greater (D. Keaffaber, personal communication, June 4, 2021).

1.1.1.4 Spiral Wound

In the spiral winding process, new pipe is installed inside the host pipe by winding a continuous strip of PVC or steel reinforced polyethylene (SRPE) material to form a pipe within a pipe. The PVC and SRPE material have

FIGURE 5.2 Geopolymer Spray Application (Reprinted with permission from Geotree)

interlocking edges that hold it together and position it inside the sewer. Annular space between the liner and host pipe is filled with grout. If the spiral liners fit flush with the inside of the host pipe, grouting may not be needed (Najafi & Gokhale, 2005). Spiral wound pipe is typically used for pipe diameters of 152 to 2743 mm (6 to 108 in.) (WEF et al., 2009).

1.1.1.5 Panels

Fiberglass reinforced plastic panels (FRPP) can be manufactured as pipe segments or panels and can be configured to fit various sizes and shapes of sewers. Fiberglass reinforced plastic panels is considered a specialty type liner that is used for larger sewer rehabilitation of circular and unusual shapes, such as egg, horseshoe, or ones with negative curvature (Najafi & Gokhale, 2005). Individual segments can be installed by hand or continuous lengths installed through pushing. Fiberglass reinforced plastic panels will match the interior dimensions of the sewer, but there is often an annular space that must be filled with structural grout. Fiberglass reinforced plastic panels can be used on pipes greater than or equal to 762 mm (30 in.) (WEF et al., 2009).

1.1.1.6 Spray-On/Spin-Cast

Spraying and spin-casting coatings onto the interior surface of the sewer is accomplished by a rotating spray head that is pulled through the pipe at a constant speed. Coatings may be epoxy resin or cement mortar based. The key to this process is ensuring the coatings are applied at a constant thickness around the circumference of the pipe and that all active infiltration has been stopped. Troweling may be needed after application to smooth out rough spots.

1.1.1.7 Shotcrete

Shotcrete is a mixture of cement and water pneumatically projected through a hose onto the surface of pipes to restore integrity. Shotcrete is typically applied to larger pipes because this requires entry into the sewer. Steel mesh or rebar reinforcement is typically installed before applying the shotcrete to the interior surface. Shotcrete is typically applied to sewers of 1219 mm (48 in.) or greater (WEF et al., 2009).

1.1.2 Grouting

Leaking pipe joints can be sealed using injection grouting. Pipe joints are tested using closed circuit television (CCTV) and a remote packer device. An air test is performed with the packer device sitting over the joint to determine if it will hold air or not. If the joint does not hold air, a low viscosity grout is injected through the defective joint and into the surrounding soil, stabilizing the void. Multiple types and variations of grout are available, depending on the site conditions, such as pipe diameter and soil type. Once the grout has cured, a second test is performed to see if the joint will hold air. If the joint does not hold air, the process is repeated: additional grout is injected, cured, and tested again. Grouting can be performed in pipes 152 to 2438 mm (6 to 96 in.) (Logiball, Inc., 2021).

1.2 Replacement

This section discusses sewer replacement methods but will not duplicate methods previously discussed in Section 1.1, Rehabilitation, that may also be used for sewer replacement. The sewer replacement methods discussed can be used for main line sewers and laterals.

1.2.1 Open Cut

Open cut is a method of construction that uses a back hoe to excavate a trench to a predetermined depth to install new pipe. Open cut methods require site and surface restoration once completed. Open cut methods are traditionally less expensive than other methods of installation and in many cases still the most widely used method for sewer replacement.

Open cut methods are not ideal for deep installations, high groundwater table, congested utility areas, and high traffic areas. Open cut methods require workers to enter the trench and requires safety procedures for confined space entry, proper trench/slope protection measures, and other required safety protocols per the Occupational Safety and Health Administration (OSHA), state, and local requirements.

1.2.1.1 Full Replacement

Full replacement of a pipe is typically done, when near to complete structural pipe failure is imminent and may result in the loss of life or harm to the environment. Full replacement is equated to replacing the entire length of pipe between manhole(s). This method, in some cases, requires the removal of the existing pipe and installation of a new pipe. In some instances, the existing pipe may be left in place and abandoned, and a new pipe installed parallel. In the latter option, all active connections are required to be disconnected and reconnected to the new pipe. This option is most economical in rural areas or areas with limited underground utilities.

1.2.1.2 Point Repair

Point repairs are typically done in emergency situations in which pipe failure has occurred or to replace large sections of existing broken/fractured pipe segments. The latter is typically referenced as a rehabilitation method. Point repairs do not require replacement of the deteriorated pipe with the same material, but proper joining of differing materials is critical in addition to using proper methods for bedding and backfilling.

1.2.2 Semi-Trenchless

Semi-trenchless is a method that requires the excavation of launching and/or retrieval pits to lower equipment into the trench for installation of the pipe using trenchless technologies. Geology, length, slope, number of service connections, and utilities are key factors to consider before using this method for the installation of a new sewer. Minimal to no disruption occurs above the ground surfaces, except at the location of the launching and retrieval pits and potential vibration during the subsurface installation. Semi-trenchless methods require workers to enter the launching/receiving pits and requires safety procedures for confined space entry, proper trench/slope protection measures, and other required safety protocols per OSHA, state, and local requirements. Several methods are discussed next.

1.2.2.1 Jack and Bore

The Jack and Bore method is a horizontal method to install pipe and requires launching and receiving pits to be excavated. A boring machine is equipped with an auger and cutting head that is lowered into the launching pit. The auger rotates the cutting head through a boring tube, as it progresses through the soil. The new pipe is pushed/pulled through behind the machine and welded to the connecting pipe. As the auger rotates, material is transported back to the machine for removal. The machine is removed at the retrieval pit, in addition to connecting the pipe to either an existing pipe or a new pipe.

Typical applications include crossing under roads, railroad tracks, small streams, or overpasses to avoid required permitting for open cut installations. This technology can also be used for installations under trees, major utilities, or environmentally sensitive areas (e.g., wetlands). The Jack and Bore method is effective for 45.7 cm (18 in.) diameter and smaller pipes.

The Jack and Bore method is best performed in stable soil with no major rocks or boulders and with a low water table (or installation above the water table). No utilities should be located at the depth of the bore. The Jack and Bore method is effective for short distances as it is difficult to maintain the required slope over long distances.

1.2.2.2 Horizontal Directional Drilling

The horizontal directional drilling (HDD) method starts with the installation of a small diameter pilot hole. A reamer is added to the drill pipe to expand the size of the hole beneath the surface. Depending on the size of the hole required, additional passes are made with increasing reamer sizes. After the hole has been expanded to the desired diameter, the new pipe is inserted and pulled through while attached to the reamer. Unlike the Jack and Bore method, the HDD method can be done over long distances. Excavations would be required to reconnect all service/lateral connections in addition to other connections to the sewer. A benefit to using HDD is the ability to avoid conflicts with other utilities as the sewer is typically installed below other utilities.

Factors that may affect the outcome of using the HDD method include subsurface and geological conditions, distance, and topography. Risks increase over longer installation lengths as a result of forces being applied to the pipe resulting in potential failures. Potential collapses of the hole and the pipe getting "stuck" or "lodged" during the pull-through process are additional potential risks.

1.2.3 Trenchless

Trenchless methods do not require the excavation of launching and/or retrieval pits. Access is through a constructed shaft. Geology, length, slope, number of service connections, and surrounding utilities are key factors to consider before using this method for the installation of a new sewer. Minimal to no disruption occurs above the ground surfaces, except at the location of the launching and retrieval pits and vibration during the subsurface installation. Trenchless methods do not typically require workers to enter the shafts as control is remote. Some installations require access within the shafts, requiring safety procedures and protocols be put in place per OSHA, state, and local requirements. Some trenchless methods require service connections to be excavated in advance of performing work on the main sewer.

1.2.3.1 Pipe Bursting

Pipe bursting replaces an existing pipe with an equal or larger diameter pipe. The process requires a winch system that pulls a pneumatic pipe bursting tool with an expander cone through the existing pipe. The pipe bursting tool breaks the existing pipe as it is pulled through the existing pipe and displaces the fractured pipe into the existing soil. The expander is located on the back of the tool and widens the old excavation while pulling in the new pipe. All service connections are required to be reinstated with this process.

Pipe bursting methods are applicable for 10 to 61 cm (4 to 24 in.) sewers in urban areas. Excavations are required to disconnect all service/lateral connections and other connections before performing pipe bursting. Once the new pipe is installed, the service/lateral connections and other connections are required to be reconnected. Pipe bursting methods are not effective on ductile iron or steel pipe, in bedrock, expansive soils, installations below the groundwater table, or areas in proximity to other utilities and structures because of the vibration of the equipment, which may cause settling.

1.2.3.2 Microtunneling

Microtunneling uses a microtunneling boring machine (MTBM), and installation uses a remote-controlled, pipe jacking method, see Figure 5.3. This method requires two vertical shafts for launching and retrieval of the equipment. Shafts are required to be sized to accept the microtunneling machine. At the face of an MTBM is a cutterhead that cuts a path through a variety of geological conditions; the type of cutterhead will be designed based on the

FIGURE 5.3 Microtunneling (Reprinted with permission from Horner & Shifrin)

soil/rock conditions, required counterbalancing of the groundwater pressure, and management of the slurry.

Microtunneling is similar to the Jack and Bore method, with microtunneling having the additional benefits of being remotely controlled, installation over longer distances, larger pipe sizes, and adaptability to a variety of soil and geologic conditions.

Microtunneling methods are not limited to restrictions as other methods such as bedrock, high water tables, unstable soil conditions, and deep installations. Typical sizes are 76.2 to 290 cm (2.5 to 9.5 ft).

2.0 LATERAL OPTIONS

A lateral is the pipe that connects a building, residential home, or business to the main line. This pipe is often the responsibility of the property owner to maintain and repair; however, some jurisdictions may own the connection to the main sewer or sections of pipes within the right-of-way. Local agencies with jurisdiction for the public system and municipalities should be consulted to determine at what point the responsibility changes from the sewer agency to the property owner.

2.1 Lateral Rehabilitation

The following sections discuss technologies used to rehabilitate the lateral pipe and connection to the main sewer. Laterals should be televised before rehabilitation to determine if point repairs are required and to locate bends or other connections in the lateral. If installing a liner, contractor installation is critical to the success of the lateral liner because of bends in the pipe resulting in wrinkles or folds in the liner. Closed circuit television inspection should be completed after installation to verify that folds or wrinkles do not generate restrictions in the flow that could result in blockages.

2.1.1 Cured-in-Place Pipe

The primary method for lateral rehabilitation uses CIPP. Section 1.1.1.1, Cured-in-Place Pipe, details the various curing methods.

2.1.2 Connections to Main

The connection to the main sewer can be a source of infiltration and inflow, root intrusion, structural damage, and odors that result in the need for improvements.

2.1.2.1 Liners

Many liner products exist to repair the lateral connection to the main sewer. These products are grouped into brim-style or full circle liners. The two types represent the amount of liner in the main sewer.

2.1.2.1.1 Brim-Style Liners

Brim-style liners cover a portion of the main sewer several inches around the lateral connection and extend into the lateral. These liners are installed from the main sewer and do not need a cleanout during construction. These can be installed before or after the main sewer has been lined. Installation in pipes that have already been lined requires any coating on the liner to be removed and cleaned. Bends in the lateral upstream of the main line can present difficulties during installation and may result in folds or wrinkles that can restrict flow.

2.1.2.1.2 Full Circle Liners

Full circle liners cover the entire pipe in the main sewer around the lateral connection and extend into the lateral. These liners generate a homogeneous material from pipe to the lateral. The length of the liner extension into the lateral can vary from 15.2 to 6096 cm (6 in. to 200 ft), based on the product. Longer extensions may require a cleanout for installation. The length of the extension can be changed based on the agency jurisdiction, condition of the lateral and distance that needs to be rehabilitated, and distance to significant bends in the lateral.

2.1.2.2 Hydrophilic O-Ring

Hydrophilic compression O-rings are used at the end of the liners to create a watertight seal at the interface between the pipe and liner at the manhole. Resins can pull away from the pipe material because of shrinkage resulting in a gap between the materials. These gaskets expand when wet to create a long-term end seal at the end of the liner.

2.1.2.3 Caulking

Hydrophilic caulk can also be installed at the end of the liner to generate a watertight seal. Caulking is fluid material and can take several days to cure. In addition, caulk does not have as much volumetric expansion as gaskets.

2.1.2.4 Grouting

Leaking lateral pipe joints can be sealed using injection grouting in the same manner as leaking pipe joints described in Section 1.1.2, Grouting, and as illustrated in Figure 5.4.

2.2 Lateral Replacement

Laterals can be replaced by traditional excavation and installation of new pipe as discussed in Section 1.2.1, Open Cut, and Section 1.2.2, Semi-Trenchless, or can be replaced using pipe bursting as discussed in Section 1.2.3.1, Pipe Bursting.

3.0 MANHOLES

Manholes may require rehabilitation or replacement.

3.1 Manhole Rehabilitation

Manhole rehabilitation is done to accomplish several factors including, reinstatement of structural strength, providing structural strength, and

FIGURE 5.4 Lateral Grouting (Reprinted with permission from Logiball, Inc.)

minimizing infiltration and inflow and corrosion. Table 5.1 summarizes the technology, and the following paragraphs provide additional details.

3.2 Manhole Replacement

Replacement of a manhole occurs when significant failure has occurred and there is a risk of collapse. Installation of a new manhole on an existing system requires the use of a prefabricated set-over manhole that will be placed over the existing sewer.

TABLE 5.1 Summary of Manhole Rehabilitation Methods

Rehabilitation Method	Reestablish Structural Strength	Provides Additional Strength	Infiltration and Inflow Control	H_2S Corrosion Protection	Pro	Con
Liners (CIPML)	x		x	x	• Can be installed as a structural chimney seal or full depth lining • Monolithic	
Inserts (full replacement within existing structure)						
• Polymer concrete (composite) composed of inert resin, sand, and rock	May		x	x	• Provides corrosion protection throughout the entire manhole vs. just the surface as provided by a coating • Requires minimal excavation and disruption to the surface • Prefabricated, lightweight	• Requires fiberglass reinforcement to reestablish structural strength • Requires removal of cone and grade rings • Loss of 15.2 cm (6 in.) of diameter
• Fiberglass	x		x	x	• Effective on brick and block manholes	• Requires removal of cone and grade rings • Loss of 15.2 cm (6 in.) of diameter
• Poured-in-place (monolithic liner system)	x		x	May		• Requires a coating or an additive to protect against corrosion

(continued)

TABLE 5.1 (*Continued*)

Rehabilitation Method	Reestablish Structural Strength	Provides Additional Strength	Infiltration and Inflow Control	H_2S Corrosion Protection	Pro	Con
Coatings (minimal surface disruptions)						
• Cementitious mortar (CM)	x		x	May	• Corrosion protection if CM contains an additive or is coated with polymer or epoxy • Inexpensive option	• Susceptible to corrosion if no additive • Inconsistent thickness during application
• Polymer and epoxy		x	x	x	• Reduced calcium hydroxide concentrations, which reduces potential for corrosion of host material	• Can be challenging to apply around steps or anything attached to the walls • Inconsistent thickness during application
• Geopolymer spray		x	x	x	• Reduced surface prep time compared to polymer/epoxy • Inexpensive • Easy application	
• Shotcrete, gunite, or sprayed concrete (composed of cement, aggregate, and water)	x		x			• Inconsistent thickness during application • Susceptible to corrosion • Requires the installation of a steel mesh or rebar before application

Grouting (injected into holes/cracks) • Walls • Bench • Connections	x	• Epoxy material • Easy method to seal • Inexpensive option	• Life of grout varies • May need to redo
Chimney seals (mechanical)	x	• Install on new or existing manholes • Made of flexible rubber/urethane • Covers vertical heights from 30.5 cm to 61 cm (12 to 24 in.)	
Inserts (placed in the top of the manhole)	x	• Various materials including stainless steel, polyethylene, and HDPE • Easy to install • Inexpensive infiltration and inflow reduction	
Grade rings	x	• Replacement can be made with a polypropylene ring • Easy to install • Inexpensive infiltration and inflow reduction	• Concrete may deteriorate over time allowing infiltration and inflow

4.0 LIFT STATIONS

Rehabilitation or replacement technologies for lift station wet wells and dry wells are similar to the manhole technologies described in Section 3.0, Manholes. Additional reference sources are listed in Section 7.0, Suggested Readings.

5.0 FORCE MAINS

Force main rehabilitation and replacement can be more complex than sewer rehabilitation and replacement because many force mains cannot be removed from service for any significant length of time.

5.1 Rehabilitation—Flexible Fabric-Reinforced Pipes

Flexible fabric-reinforced pipes (FFRP) is pipe that is winched into the host pipe and then inflated to fit against the inside wall of the pressure pipe. Flexible fabric-reinforced pipes is a fabric that is sandwiched between two layers of thin-walled plastic pipe, such as polyethylene (PE) and HDPE. When installed, the FFRP is independent of the host pipe. Bends and fittings are accommodated by this process for various pressure conditions. Long lengths can be installed because the pipe is delivered to the site on a reel. Connections made to existing pipe are either welded or use flanged end fittings. Flexible fabric-reinforced pipes is limited to smaller diameters.

5.2 Replacement

Force main replacement methods are similar to those discussed under Section 1.2, Replacement. Factors to consider are flow bypassing, pipe material, access points, soil condition, groundwater, and depth.

6.0 REFERENCES

Logiball, Inc. (2021, May). https://www.logiball.com/products/test-and-seal-packers/

Najafi, M., & Gokhale, S. (2005). *Trenchless technology—Pipeline and utility design, construction, and renewal.* McGraw-Hill.

Water Environment Federation, American Society of Civil Engineers, & Environmental & Water Resources Institute. (2009). *Existing sewer evaluation and rehabilitation* (3rd ed., WEF Manual of Practice No. FD-6, ASCE Manuals and Reports on Engineering Practice No. 62). Water Environment Federation.

7.0 SUGGESTED READINGS

American Society of Civil Engineers. (2001). *Standard construction guidelines for microtunneling.*

Deboda, T., Giokas, K., Harmer, M., Morris, W., Nelson, J., Sever, F., Sumner, T., & Tobey, B. (2021). *Fact sheet—Sanitary manhole rehabilitation* (draft). Water Environment Federation.

Deboda, T., Jurgens, D., Lehman, R., Rigby, D., Osborn, L., Sumner, T., & Vallabhaneni, S. (2017). *Fact sheet—Sanitary sewer rehabilitation* (WSEC-2017-FS-009). Water Environment Federation, Collection Systems Committee. https://www.wef.org/globalassets/assets-wef/direct-download-library/public/03---resources/wsec-2017-fs-009---csc---sewer-rehabilitation---final---9.27.17.pdf

Hubbard, P., Barsanti, J., Shaw, J., & Zheng, X. (2019). *Fact sheet—Sanitary sewer systems: Lift stations and data management* (WSEC-2019-FS-013). Water Environment Federation, Collection Systems Committee. https://www.wef.org/globalassets/assets-wef/direct-download-library/public/03---resources/wsec-2019-fs-013---csc-mrrdc---lift-stations-and-data-management---final.pdf

National Association of Sewer Service Companies Lateral Committee. (2012). *Overview of lateral and main/lateral connection lining and sealing technologies.* National Association of Sewer Service Companies.

U.S. Environmental Protection Agency. (2006). *Water technology fact sheet—Pipe bursting* (EPA 832-F-06-030). Office of Water.

Water Environment Federation & National Association of Sewer Service Companies. (2019). *Trenchless technologies and asset management* (3rd ed., WEF Manual of Practice). Water Environment Federation.

6

Intelligent Wastewater Collection Systems

Susan Moisio, PE; Anjulie Cheema; Reese Johnson, PE, PMP; Daniel J. Murray Jr., PE, BCEE; & Matthew Van Doren, PE

1.0 CHARACTERISTICS OF AN INTELLIGENT WASTEWATER COLLECTION SYSTEM

Wastewater utilities face increasing challenges to operate and maintain their collection systems more effectively to ensure capacity, performance, and resiliency. Over recent decades, innovations in condition assessment and rehabilitation technologies and approaches have underpinned the emergence and successful application of strategic asset management to meet these challenges. More recently, innovations in advanced sensing technologies; ubiquitous and increasingly reliable communication systems; advanced data management and analysis tools; and state-of-the-art machine learning methods have become the foundation of a transformational approach to managing collection systems. "Real-time control and optimization" is this new paradigm for the operation, maintenance, and management of wastewater collection systems (WCSs). The emergence of this approach has led to a new moniker for these intelligent systems: "smart sewers."

Although the driving forces behind the growing development and implementation of smart sewers varies from utility to utility, there are common elements that distinguish these as "smart" sewer systems. This chapter presents the characteristics and components of a smart sewer system, along with discussions of the data management and analytics, cloud computing, and cyber-security challenges and issues that go along with them. This chapter also presents an evaluation of changing workforce needs, including the skills and continuous learning required to take advantage of new technologies and achieve the potential of truly intelligent wastewater collection.

1.1 The Smart Water Network (Modified for Wastewater)

The text used in this chapter has been adapted to wastewater with the permission of the Smart Water Networks (SWAN) Forum (SWAN Forum, 2021). An intelligent WCS, a specific application of a smart water network applied to wastewater collection, has the ability to operate beyond an autonomic response to stimuli. Such a system goes beyond allowing passive forces, like gravity or pressure gradients (provided by pumps), to control the flow of wastewater. An intelligent WCS collects data and turns the data into information that make its actions "smarter." Further, it can uncover patterns in that information that illuminate knowledge, allowing it to develop rules and change future actions based on predicted conditions. An intelligent WCS has certain components. To understand these components and how they interrelate, one must understand the layers of a smart water network as illustrated in Figure 6.1.

Pipes, diversion chambers, gates, constructed outfalls, flow control mechanisms, pumps, and pump stations are all elements of the *physical*

FIGURE 6.1 The Five Layers of a Smart Water Network (Adapted from and reprinted with permission from SWAN Forum)

layer of a smart water network. As its name suggests, this layer consists of the physical elements enabling the collection and delivery of wastewater to the end points of the system. These are data-less elements that typically perform mechanical, hydraulic, or chemical functions. Although the physical layer does not have data interfaces, it can be controlled using data collected in the next sensing and control layer. Although there may be valuable innovation and design in the physical layer, a system that is composed of only the physical layer is not an intelligent WCS.

The *sensing and control layer* is composed of equipment and sensors that measure parameters of the wastewater collection and delivery (such as level, flow, pressure, and water quality parameters) and remote-controlled devices enabling remote operation of the network (such as remote-controllable pumps, valves, gates, and weirs). In essence, the sensing and control layer is the only interface between the network operator's data systems on one side, and the physical layer on the other side, enabling the connection of the "smarts" of the intelligent WCS to the physical network. Elements of this layer typically have one "wet" end or aspect with direct contact or relation to the wastewater (such as a valve or the mechanical end of a flow sensor), and one "dry" data interface (such as a valve controller input, or a sensor's data output).

The *collection and communications layer* is responsible for discrete data point collection, transmission, and storage. Through two-way communication channels, commands sent to the second layer instruct sensors and actuators what data to collect or which actions to execute. For example, a fixed cable network, radio, cellular, Wi-Fi, and other communication technologies related to data transfer are all part of this layer. This is the first completely "dry" layer because it moves data only between the sensing and control layer and the higher layers.

The *data management and display layer* is where data from different sources comes together and may be used by operators. Data may be pre-processed, stored, transferred, and accessed by central systems. This is where human operator commands or instructions from higher-level systems are interpreted as device settings (such as an operator setting a facility into "wet weather mode," which becomes discrete commands that change gate positions and pump settings, etc.). This interfaces with the underlying communications infrastructure on one side, and with a human operator or other central data systems on the other side. The graphical user interface (GUI) provided by many supervisory control and data acquisition (SCADA) systems (or developed in-house) often falls into this layer; such GUIs have some data validation capabilities but are primarily focused on the display of multiple data streams graphically and in context. Other components in this layer include data repositories, geographic information system (GIS) or network schematic visualization tools, control systems with simple alert rules, and fixed-rule feedback automation (e.g., adjusting a gate to meet a desired flowrate).

The *data fusion and analysis layer* brings together raw input data and derives processed knowledge that was not previously discernable from the data as collected. The resulting information may be displayed to a human operator, be passed on for further analysis within the layer, or trigger automatic action by means of the data management layer (or directly via the communications layer). Data from multiple readings, data sources, and even data types become high value information, in the form of alerts of problems, automated responses to system changes, high-level summaries, or forecasts of future system conditions. Components in this layer may include hydraulic modeling systems, condition assessment tools, flow optimization strategies, dynamic (not fixed feedback) pumping algorithms, and decision support systems. This layer in a WCS is the epitome of an intelligent WCS.

1.2 Actionable Data

Data are the lifeblood of an intelligent WCS. Without data, there is no need for layers two through five of the network: the layers that turn data into information and further into knowledge have nothing to act upon. However, that data must be *actionable* if the intelligent WCS is to be effective. Actionable data has four components: type, frequency of collection, timeliness, and accuracy.

Type: What parameter is being measured? Is it level only or is it flow (with both a level and velocity component)? If an operator is trying to avoid an overflow from a WCS, level is often sufficient; however, if the operator is trying to optimize a gate position to control flow in a system that experiences backwater conditions, velocity is a necessary component. The type of data are dependent upon choosing the correct measuring device or meter (an element of the sensing and control layer) for the parameter that is desired. The technologies for measuring various parameters are identified in Chapters 2 through 5 in this publication.

Frequency of collection: How frequently is the parameter being measured? Is it every minute or every hour? If the system conditions change rapidly, such as in a collection system with stormwater inputs from a highly urbanized area, data collection must be more frequent, likely on the scale of minutes. If the system is less dynamic, such as at an equalization basin with a large surface area, data collection rates can be less frequent, on the scale of every 15 minutes or once an hour. Ultimately, data collection must be frequent enough that changes in the system can be detected before the condition changes sufficiently again.

The frequency of data collection is a collection and communication layer configuration. An advancement has been the integration of processing capability with sensors (known as "edge computing"). Although a basic collection and communication layer collects measurements at static intervals (e.g., every 5 minutes), the more advanced layers dynamically change the collection (and transmission) frequency based on the measured conditions. For example, the Metropolitan Sewer District of Greater Cincinnati (Cincinnati MSD) uses an advanced collection and communication layer in which the collection system level sensors collect the water level every 5 minutes but transmit the accumulated data only once an hour under normal flow conditions. If the level is above a set threshold, or changes rapidly, the transmission interval is reduced, and data are transmitted every time the sensors take a reading (once every 5 minutes).

Timeliness: How quickly are data available to the decision maker, whether that decision maker is a human operator and an element of the data management or data analysis layers? Can the decision to close a gate be made within seconds of an elevated level being detected at a manhole, or will the decision take minutes or hours? Timeliness depends on the speed of the collection and communication layer as well as the ability of the data management and data analysis layers to store, retrieve, and display the data from the sensor device layer. Timeliness is also affected by the need to process the raw data, which includes transforming the measurement into the correct units but most importantly checking its quality. Often timeliness must be balanced with accuracy: taking time to check the data may result in the data not being available in time to affect the performance of the system, rendering the data useless. For example, an operator needs to see that a level in a wet well is high and continuing to rise unabated to know that an additional pump is necessary quickly, before an overflow occurs or the pump station is flooded.

Accuracy: How closely does the data represent the actual value of the metric at the time measured? Raw data can be noisy, and communication systems can experience faults, so inaccurate, erroneous, and missing data must be expected. However, the operators must be able to trust the system, if the intelligent WCS is to be effective. Both the data management and the data analysis layers can include data validation tools that help improve the quality of the data, and thereby build trust in the system. Simple tools exist that ignore values outside a specific range, require multiple data points that exceed a static limit before acting, or smooth the data stream by taking a rolling average of the data, among others. These simple tools have the advantage of being fast and readily implementable using commercially available products such as SCADA software. More advanced data validation tools use numerical models or various forms of machine learning to develop (predict) an expected value for a parameter, and then compare it to the measured value. Values too far away from the predicted value are ignored. The sooner the decision maker needs the data, the more automated these processes must be.

Ultimately, actionable data must be "fit for purpose." Fit for purpose means that a "good enough" measure of the correct metric is available frequently and quickly enough for a decision to be made. At Cincinnati MSD, all known overflow points are monitored remotely by electronic level sensors providing data to the watershed SCADA system every hour during normal flow conditions. If levels rise above the maximum value for dry weather conditions, the SCADA system receives data every 5 minutes, and depending upon the length of time the level rises above the overflow level, different actions are triggered. For a single time step, the data stream is flagged for later review, while for two or more time steps, maintenance personnel are notified immediately to respond and mitigate the effects of the discharge. Regardless of the response outcome, the data stream is automatically extracted each night, outliers are removed, and null values populated, and then run through a hydraulic model to calculate overflow volume. If the overflow event is determined to be a reportable event, a data analyst is automatically notified, and the operator reviews the overflow volume and level time series before submitting a notice to the regulatory agency within a required 24-hour window.

Once a system has actionable data, it needs to have assets that can be manipulated and controlled by that data to affect system performance. The next section explains what a controllable asset is, and the range of ways it can be controlled.

1.3 Controllable Assets

A necessary characteristic of an intelligent WCS is the dynamic control of assets in real time in response to system conditions. These controllable assets inherently must be capable of adjustment, such as pumps, gates, and certain weirs. Implementing dynamic control provides utilities an opportunity to optimize both existing and proposed assets to meet operational objectives and regulatory requirements effectively and efficiently.

To dynamically control an asset, it must be connected with a real-time control (RTC) system. An RTC system, at a minimum, includes sensors that measure the process, control elements that adjust the process, and data communication between them (Döring et al., 1989). A controllable asset is connected to the inputs of a local RTC device, such as a programmable logic controller (PLC) or remote telemetry unit (RTU). The RTC device provides outputs that adjust the controllable asset based on the difference between specified set points and measured variables. These set points and measured variables can be defined locally or from a remote system using a SCADA system.

A SCADA system is control system architecture that uses computers, networked data communications, and a GUI for high-level process supervisory management. A SCADA system is a primary component of an RTC system, communicating data and control commands between the network of sensors and controllable devices in an intelligent WCS to the central systems that process, store, and archive the data. Additionally, a SCADA system includes GUI tools that allow for monitoring and manual control of remotely located assets.

Three distinct levels of control can be implemented in an RTC system: local, regional, and global. These levels of control are generally classified by the extent of their effect on the WCS. The higher levels of control require more data, more complex control logic, and a greater reliance on the data collection and communication layer, but also achieve a greater level of optimization and reach higher performance goals.

Local control is the simplest form of automatic control used to manage isolated capacity issues that require only information collected near the remote site. An example of local control is the automatic adjustment of a gate to maintain a specific flowrate based on measurements from a flow meter located in the downstream sewer. This level of control is appropriate where the controllable asset does not affect other parts of the WCS or where information from other sites is not available. Although it can effectively optimize the singular asset, it does not account for conditions elsewhere in the WCS.

Regional control achieves greater overall performance in the WCS by incorporating data from multiple sites. This level of control requires data to be communicated between multiple remote sites and involves more complicated control logic. This level of control can be applied where there are multiple points of control in the WCS feeding an asset with a fixed conveyance capacity, like an interceptor, or where a single facility such as a pump station has an effect on much of the WCS.

Global control achieves the greatest level of performance from the WCS by coordinating the control actions at numerous remote sites and accounting for the variability inherent to input factors such as predicted precipitation and time of travel within the system. Such operational control strategies are typically developed through computerized optimization techniques and are refreshed at intervals no longer than 15 minutes.

Cincinnati MSD has implemented an intelligent WCS that has progressed through all three levels of RTC and now uses them all in concert. Initially, local control was implemented at each of its remote wet weather facilities, including three combined sewer overflows retrofitted with inflatable dams, pictured in Figure 6.2, to maximize in-system storage. Although this remains the fundamental control strategy for these assets, once its watershed SCADA system linked the remote facilities and collection system sensors in a shared communication and control network, Cincinnati MSD achieved regional control of these assets. At this stage, the facilities were able to optimize their operations based on relevant sensor data from across the basin. For example, a remote chemically enhanced high-rate treatment (CEHRT) facility in Cincinnati MSD's collection system was tied into the watershed SCADA system and then operated autonomously based on data from level sensors strategically located at hydraulic choke points downstream and along feeder pipes upstream that anticipate what flows will reach the facility. Finally, once agent-based trading rules were added to the PLCs at each remote facility, the inflatable dams and the remote high-rate treatment facility worked together to fully utilize the available capacity in their shared downstream interceptor sewer during wet weather. This has achieved a level of global control that coordinates all of the controllable assets in the basin to maximize conveyance to the water resource recovery facility (WRRF) and minimize overflows across the entire basin.

FIGURE 6.2 Inflatable Dam (Reprinted with permission from the Metropolitan Sewer District of Greater Cincinnati)

1.4 Visualization

Data visualization is the presentation of data, raw or processed, in a graphical format for human users of the system. It is a key component in harnessing the large amount of data collected from the network of sensors and other sources within a WCS and transforming it into information and insight that can be used in the operation of the system. Data visualization is most effective when it provides that data in context with other data streams or physical attributes of the system. For example, having a time series of level measurements from a manhole is helpful, but seeing that manhole's level on a map of the system with other level readings enables an operator to see trends.

It is important to ensure that the data type and analysis objectively inform the development of a specific data visualization. Trying to fit data to a specific visualization type will not effectively convey information. Simple, focused data visualizations that are tailored to the analysis objectives of the end user bring together different data sources to build a compelling story that empowers end users, from operators to managers, to make good decisions and take effective action.

When developing a data visualization, it is important to consider the devices that will be used to display the visualization. Desktop and web applications can be secure methods for connecting end users to data visualizations. Desktop applications have the benefit of generally being more easily developed by internal resources, but inherently limit the number of individuals that can access the visualization. The development and maintenance of web applications generally require technical resources (a user interface developer, database manager, functional programmer, etc.) that a utility may not have internally available. However, a web application allows many end users to access the data visualization simultaneously.

Cincinnati MSD has developed data visualizations used by operators, maintenance personnel, and managers across departments. Using a cloud-based web application, end users can access the data visualizations through multiple platforms at any time. Maintenance personnel can review equipment status and the performance of remote unmanned facilities. By exploring the visual trends, personnel can proactively identify and respond to issues. Additionally, treatment facility operations at Cincinnati MSD's large WRRFs have been improved by incorporating analytics into a data visualization that projects future flow to the facility, as illustrated in Figure 6.3.

1.5 Analytics

Finally, analytical capabilities are the characteristic of an intelligent WCS that transform data into information (and further into knowledge). With or

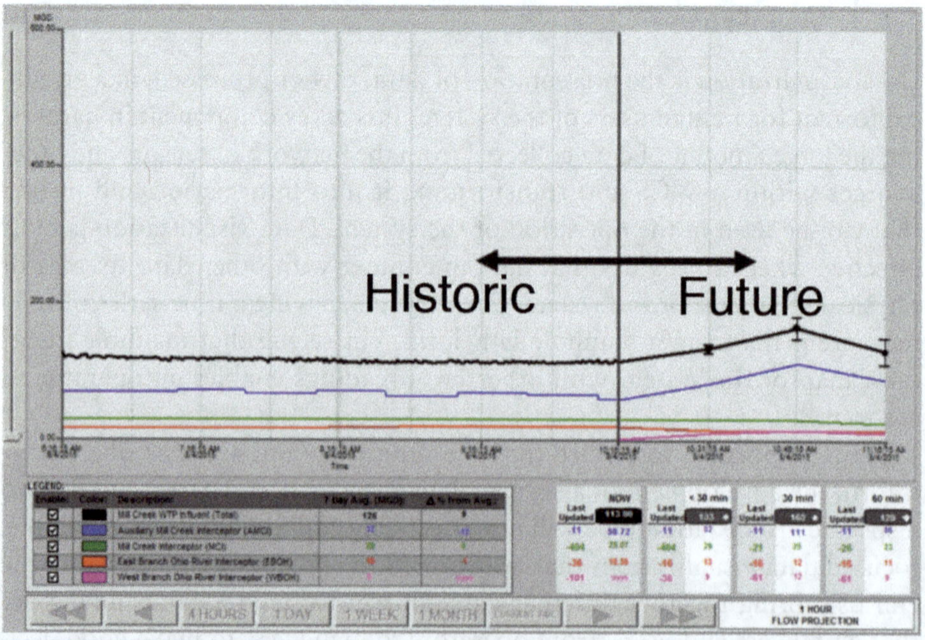

FIGURE 6.3 Cincinnati MSD Data Visualization (Reprinted with permission from the Metropolitan Sewer District of Greater Cincinnati)

without visualizations, analytical tools can exist in the collection and communication layer, through the deployment of "smart sensors" that imbed PLCs with a measuring device and leverage "edge computing" to make some decisions at the site, but typically exist in the monitoring or control (i.e., SCADA) software that also manages and displays the data. Analytics are the pre-frontal cortex of the system and can be deployed at varying levels of sophistication.

A common analytical tool that can be leveraged by an intelligent WCS is a computer model of the collection system. Whether is it simply a hydraulic model, or a more complex model that includes water quality parameters, or an integrated model with a surface flow representation, computer models can be leveraged to gain insight from the vast data collected in an intelligent WCS. For example, direct measurement of overflow volumes from a WCS is difficult. To overcome this limitation and still meet regulatory requirements of a National Pollutant Discharge Elimination System (NPDES) permit to report this metric, Cincinnati MSD uses the level and flow data collected both from the collection system as well as the receiving streams to feed a hydraulic model of the basin that runs autonomously every night. The overflow occurrence and volume data are available the next day for review, analysis, and, most critically, on-time reporting to the regulatory agency. The results are aggregated leveraging a Microsoft Power BI dashboard, as shown on Figure 6.4, to see trends and uncover anomalies, which are reviewed weekly with operators to identify changes that can improve performance (lower overflow volumes) in the future. For example, automated alerts from the SCADA system were created based on gate positions that exacerbated overflows, and maintenance activities on pumps were increased where performance had decreased but had not yet reached mechanical failure.

More advanced analytical tools that can be part of an intelligent WCS harness machine learning to mine greater insights from the collected data. Machine learning uses a subset of the data to understand the connections between cause and effect in a system and then develop rules that model system behavior. For example, an artificial neural network (ANN) model of a WCS is trained using machine learning and then can be used to estimate future overflows based on predicted rainfall. Because the ANN model can run much more quickly than a hydraulic model, it can be fed by data directly from the system and predict conditions in near real time. When these predictions are presented alongside the actual field data, the benefits are two-fold: first, operator can immediately see where problems are developing in the collection system and react more quickly; second, the data analyst can see when the model is not accurately predicting future performance and either retrain (recalibrate) the model or recommend system investigation.

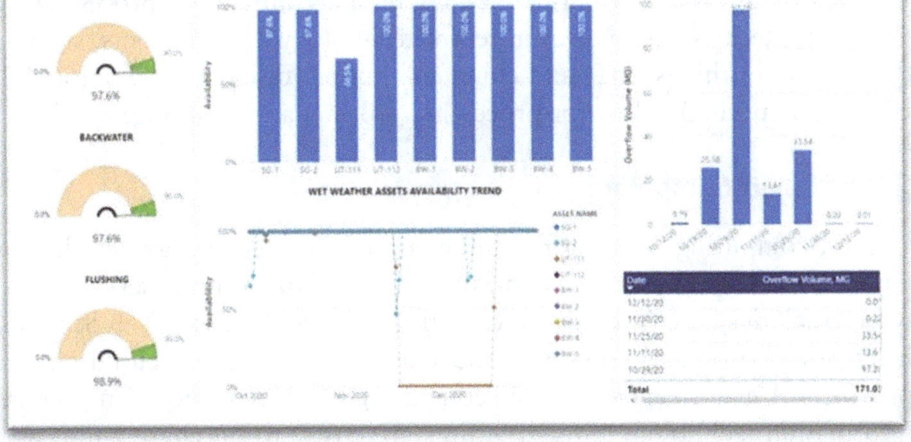

FIGURE 6.4 Cincinnati MSD Microsoft Power BI Dashboard (Reprinted with permission from the Metropolitan Sewer District of Greater Cincinnati)

2.0 ROLE OF BIG DATA, CLOUD COMPUTING, AND THE INTERNET

Utilities are becoming overwhelmed with data. For example, water and wastewater utilities manage infrastructure asset data, customer information, SCADA data, sensor monitoring data, and more. A 2018 Water Research Foundation (WRF) report queried chief information officers from 50 of the largest water and wastewater utilities in the United States on the amount of data they use of the data they generate (WRF, 2018). Only 10% is actually used, mostly because of a lack of big data analytics tools and staff resources. The unused data (often referred to as dark data) could provide insights into ways to improve operational efficiency and enhance regulatory compliance. Harnessing big data using analytics could provide utilities with a variety of benefits.

2.1 Big Data

Big data refers to the increasing amount and variety of data, and has three defining principles:

- Volume, or the large amount of data;
- Velocity, or how fast the data are generated, collected, and processed; and
- Variety, or the types of data and format (structured, unstructured, semi-structured).

Because of these three principles, big data are difficult to process using traditional techniques. However, recent technology developments, such as the internet of things (IoT) and cloud computing (discussed in Section 2.2, Cloud Computing), allow for processing, storing, and analyzing big data.

2.2 Cloud Computing

"Cloud computing" is a term referencing the on-demand availability of computer resources, such as computer power, data storage, servers, software, analytics, and intelligence over the internet. Amazon Web Services, Microsoft Azure, and Google Cloud Platform are popular cloud service providers. A major benefit of cloud computing is cost reduction. On-premise information technology infrastructures are costly to purchase and costly to maintain. Cloud computing solves this by allowing a user/organization access to the latest generation of fast and efficient computing resources and pay only for their usage. Other advantages include improved scalability, security, reliability, and performance.

The on-demand availability of an expansive and higher-quality network of servers and computing power has many uses for a collection system. One example of using cloud infrastructure is running hydraulic models to understand system performance under various alternative conditions (different infrastructure, operations, or storm events). Figure 6.5 shows a simulation management setup that allows a user to seamlessly launch a large batch of collection system model runs onto a limited number of on-site computers owned by the user or to an unlimited number of virtual machines in the cloud, perform the model simulations, and download the results. Model runs can be completed in parallel and critical results provided in a shorter time frame. This process was implemented recently for the San Diego International Airport to size a stormwater cistern (San Diego County Regional Airport Authority, 2020). More than 2000 design configurations were considered and modeled, with each model run taking 5.5 hours to complete. Instead of using four local computers, the simulations were run in the cloud and took approximately 6 days to complete, providing a time savings of

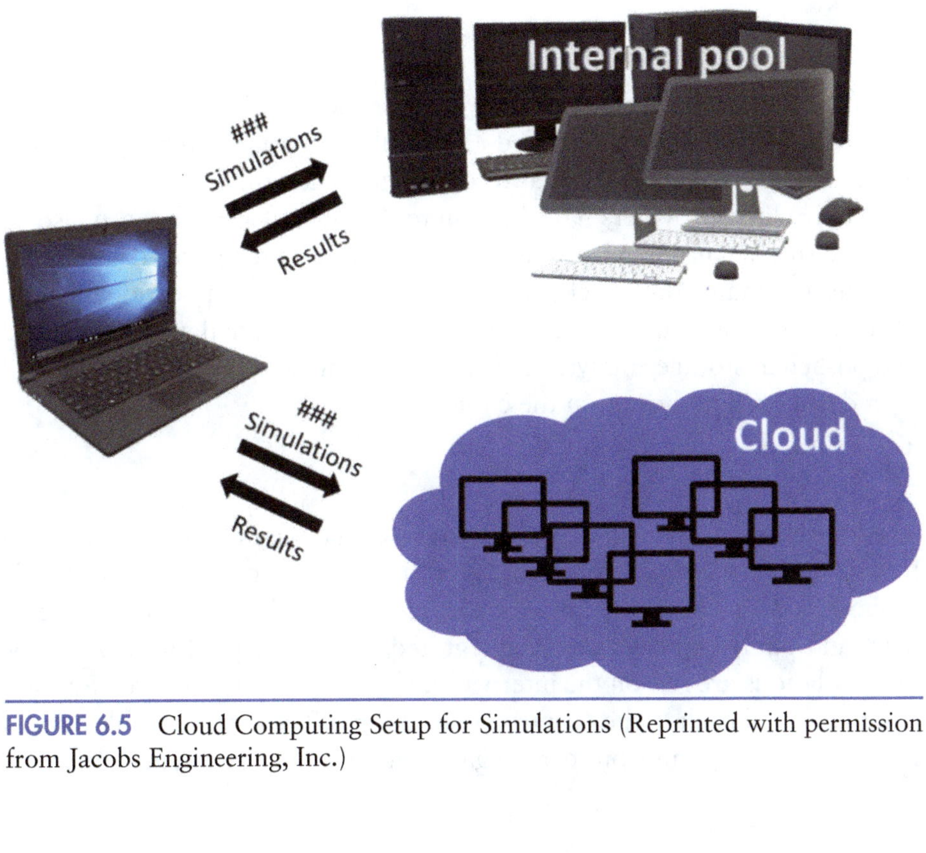

FIGURE 6.5 Cloud Computing Setup for Simulations (Reprinted with permission from Jacobs Engineering, Inc.)

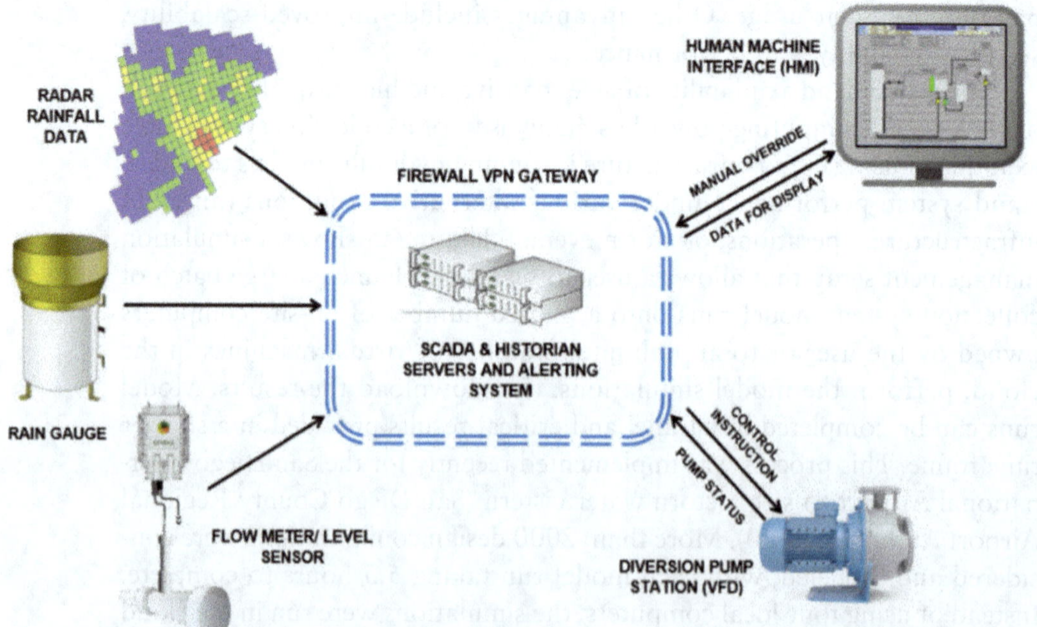

FIGURE 6.6 IoT Sample of Rainfall, Flow Data, Pump Station, and SCADA (Reprinted with permission from Jacobs Engineering, Inc.)

120 days. By evaluating thousands of possible solutions, the most cost-effective cistern size was identified and resulted in approximately $14 million in capital savings compared to the original design.

Many cloud providers have built-in tools and software that can also be applied, such as machine learning models, business intelligence dashboards, storage, automated data backup, and so on. A common application of many of the cloud tools for a sanitary system is the IoT, where data are collected and pushed to storage, analytical processes are run, and results are presented on a dashboard all hosted in the cloud.

2.3 Internet of Things

The IoT refers to a system of internet-connected computing devices enabled to send and receive data. An example of the IoT related to intelligent WCSs is shown on Figure 6.6. Rainfall data, radar rainfall data, flow and level data, and pump station data are integrated in a server platform. Decisions or calculations are run on the internet platform and control instructions are communicated back to the "controllable" devices. These controllable devices in the figure are pumps but can be gates, pumps, inflatable dams, or weirs.

The calculations can also include using rainfall data and predictive analytics to develop an understanding of future rainfall and make operational decisions accordingly.

3.0 WORKFORCE NEEDS

As wastewater utilities transform into digital, data-driven organizations and as intelligent WCSs become more mainstream, utilities face new workforce challenges. The effective management, operation, and maintenance of these smart sewer assets will require personnel with skills relating to innovative sensor systems, digital communications, and advanced data analytics. To attract the skills needed for this digital evolution, wastewater utilities must face the realities of the changing workforce.

In its report, "Building Workforce Skills for Intelligent Water Operations," the WRF identified two important dynamics that will likely affect the acquisition of workforce skills needed for the design and implementation of intelligent water systems: (a) the new skills required for the future workforce and (b) the changing demographics of the workforce (WRF, 2019).

As digital technologies become more integrated into wastewater collection and treatment systems, operators and technicians will require new and enhanced capabilities and skills. These will include managing and interpreting data, including assessing analytical outputs via system dashboards and machine learning algorithms. Skilled personnel at a data-driven wastewater utility include:

- Data managers who define and organize input from multiple data sources to certify data security and quality, coordinate data analyses and availability within the organization, and ensure the reliability and resiliency of data management systems.
- System operators who evaluate multiple data streams and system status information to support operational decision-making field verify data quality using predefined metrics and criteria to identify potential data collection problems and provide situational awareness to judge the adequacy of automated decision-making.
- Instrumentation technicians who install and monitor coordinated sensor networks to confirm equipment reliability and security, provide needed repair and maintenance of sensors and communication components, and maintain knowledge of technology advancements to upgrade and replace equipment when appropriate.

Challenges exist for wastewater utilities that need personnel with these skills, knowledge, and capabilities. The pool of qualified candidates can be limited by the competition from other, often higher paying, industries and by the perception that wastewater utilities are not an employer of choice when compared to industries with reputations for innovation and growth.

Another factor that affects the ability of wastewater utilities to hire and retain digital data and technology employees is the changing demographics of the workforce. It is estimated that by 2025, more than 75% of the workforce will be composed of millennials, those people born between 1982 and 1995 (WRF, 2019). Most millennials have grown up with mobile and readily accessible technology. Their preferences for a work environment include collaboration, instant gratification, and individuality. These workers place a high value on work-life balance and are open to job mobility, changing jobs approximately every 3 to 5 years.

The differences between generational workforce demographics require creative hiring practices and training programs. For example, the emerging workforce is more comfortable using digital technologies and expects trustworthy information to be available in real time. The more traditional wastewater utility workforce, established by the baby boomer generation, use intuition based on job experience to make operational decisions. As these and other workforce demographics change, wastewater utilities will need to balance the organization's need for skilled and effective employees with the hopes and desires of their diverse workforce.

4.0 SECURITY OF THE INTELLIGENT WASTEWATER COLLECTION SYSTEM

Security of the intelligent WCS is important for operational continuity. A disruption in service or an unplanned overflow can lead to regulatory consequences for the utility. Operational security refers to the practices and processes that are deployed to protect physical devices. Operators should evaluate the location of the devices, the sensitive nature of the data, and the redundancy of the system when deciding how much physical security is adequate. Physical security should include the securing of these areas with locked access and the use of intrusion alarms: network equipment, server equipment, control rooms, PLCs, and radio or cell phone communications.

Networks should be segregated by creating security zones. These zones allow for traffic on the networks to be inspected and monitored. The National Institute of Standards and Technology Special Publication Guide to Industrial Control Systems (ICS) Security states, "Network segmentation

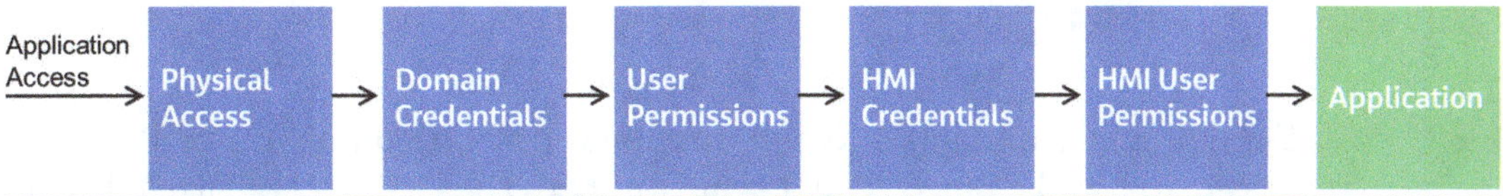

FIGURE 6.7 Concept of Defense in Depth (Reprinted with permission from Jacobs Engineering, Inc.)

and segregation is one of the most effective architectural concepts that an organization can implement to protect its ICS" (Stouffer et al., 2015).

Human-machine interface (HMI) security should include security at the workstation level, with no automatic logons and no generic accounts allowed, unique usernames required for each user, and HMI authentication should not be forced. The concept of defense in depth is shown in Figure 6.7.

4.1 Redundancy

Network redundancy improves reliability and ensures the availability of the network to operate when needed. If one system fails, the redundant system can take over, and the intelligent WCS can continue to operate.

4.2 Failover

The term failover refers to the ability of a system to switch to a backup seamlessly when needed. This is a means of protecting network systems from failure by ensuring that the backup equipment automatically takes over, typically by means of a router, when the main systems fail.

4.3 Risk Profile

Online and connected systems like intelligent WCSs face security risks; however, the controllable assets connected through an intelligent WCS can be programmed with an awareness of the risks. For example, Cincinnati MSD's remote wet weather facilities are tied together through an internet-facing control system. When fully connected, the remote facilities coordinate their control actions using data from other sites and achieve optimal performance. If communication is lost, or if the system is compromised, each remote facility defaults to local control logic, which takes inputs only from hard-wired sensors. Further, if the remote site were to experience a local failure, be it loss of power, mechanical failure, or control issue, the structure is designed to "fail open," such that sewer backups and other effects on the WCS do not occur. These steps do not eliminate the risk of compromise, and a robust cyber-security approach is necessary, but it minimizes the consequences of such a compromise.

5.0 REFERENCES

Döring, R., Schilling, W., & International Association on Water Pollution Research and Control Task Group on Real-Time Control of Urban Drainage Systems. (1989). *Real-time control of urban drainage systems: The state-of-the-art*. International Association on Water Pollution Research and Control.

San Diego County Regional Airport Authority. (2020). *San Diego County Regional Airport Authority receives environmental excellence award: Industrial Environmental Association recognizes SAN for northside stormwater capture cistern news release*. https://www.san.org/news/news-detail/san-diego-county-regional-airport-authority-receives-environmental-excellence-award

Stouffer, K., Pillitteri, V., Lightman, S., Abrams, M., & Hahn, A. (2015). *Guide to industrial control systems (ICS) security* (NIST Special Publication 800-82, Revision 2). U.S. Department of Commerce, National Institute of Standards and Technology.

The Smart Water Networks Forum. (2021). *A layered view of smart water networks*. https://www.swan-forum.com/swan-tools/a-layered-view/

Water Research Foundation. (2018). *Leveraging other industries—Big data management (phase I)* (Project No. 4836).

Water Research Foundation. (2019). *Building workforce skills for intelligent water operations* (Project No. 4663).

6.0 SUGGESTED READINGS

Environmental Resilience Institute. (n.d.). *South Bend, Indiana uses smart sewer technology to monitor and manage increased water levels*. https://eri.iu.edu/erit/case-studies/south-bend-indiana-uses-smart-technology-to-monitor-and-regulate-wastewater-levels.html

U.S. Environmental Protection Agency. (2021). *Smart data infrastructure for wet weather control and decision support* (EPA 830-B-17-004). Office of Wastewater Management. https://www.epa.gov/sites/production/files/2018-08/documents/smart_data_infrastructure_for_wet_weather_control_and_decision_support_-_final_-_august_2018.pdf

U.S. Environmental Protection Agency Region 1. (n.d.). *Wastewater collection system toolbox*. https://www3.epa.gov/region1/sso/toolbox.html

Water Research Foundation. (2019). *Integrated management of sensor data for real time decision making* (Project No. 4759).

Index

www.ingramcontent.com/pod-product-compliance
Lightning Source LLC
LaVergne TN
LVHW060559240526
839643LV00003B/47
* 9 7 8 1 5 7 2 7 8 4 2 4 6 *